195

THE SURVEY

OF

WESTERN PALESTINE.

M E M O I R

ON THE

PHYSICAL GEOLOGY AND GEOGRAPHY

OF

ARABIA PETRÆA, PALESTINE, AND ADJOINING DISTRICTS.

WITH

SPECIAL REFERENCE TO THE MODE OF FORMATION OF THE
JORDAN-ARABAH DEPRESSION AND THE DEAD SEA.

BY

EDWARD HULL, LL.D., F.R.S., F.G.S.,

DIRECTOR OF THE GEOLOGICAL SURVEY OF IRELAND, AND PROFESSOR OF GEOLOGY IN THE ROYAL COLLEGE OF
SCIENCE, DUBLIN ; HON. MEMBER OF THE GEOLOGICAL SOCIETIES OF BELGIUM, EDINBURGH,
GLASGOW, AND DUDLEY ; OF THE YORKSHIRE PHILOSOPHICAL SOCIETY, AND
THE ACADEMY OF SCIENCE, PHILADELPHIA.

PUBLISHED FOR

THE COMMITTEE OF THE PALESTINE EXPLORATION FUND

1, ADAM STREET, ADELPHI.

1886.

PREFACE.

I HAVE to express my thanks to Colonel Sir Charles Wilson for his kindness in perusing this Memoir when passing through the press, and for several suggestions. Also, to Mr. Rudler, Curator of the Museum of Practical Geology, for the notes on the microscopic structure of some rock-specimens brought home from Arabia Petræa. Professor Sollas, of Trinity College, Dublin, has also rendered much service by determining the fossils collected from the limestone of Wâdy Nasb, and some other localities ; as well as for some suggestions on biological matters, for which my thanks are now tendered. I have also to acknowledge my obligations to Dr. Günther and Mr. Eager A. Smith for determining the species of shells and corals from the raised sea-bed at Moses' Wells, near Suez.

The Memoir has been written between the intervals of official, and numerous other, engagements ; yet, the deep interest which the subject of the physical history of this region has excited in my mind has proved of itself a sufficient incentive to its study and attempted elucidation.

E. H.

DUBLIN, *December,* 1885.

212468

PREFACE.

[...] [...] [...] thanks to Colonel Sir Charles Wilson for his kind-
[...] [...] the Memoir when passing through the press, and for
[...] [...] [...] to Mr. [...], Curator of the Museum of
[...] [...] for the notes on the microscopic structure of some
[...] [...] some from Arabia Petræa[...] -Professor Sollas, of
Trinity College, Dublin, has also rendered much service by determining
[...] [...] from the limestone of Wady Nasb, and some other
[...] as well as for some suggestions on biological matters for which
[...] [...] now indebted. I have also to acknowledge my obligation to
Dr. [...] and Dr. Edgar A. Smith for determining the species
[...] [...] from the raised sea-bed at Moses' Wells, near Suez.

The Memoir has been written between the intervals of official and
other like engagements; yet the deep interest which the subject of
[...] [...] this region has excited in my mind has proved of
[...] [...] incentive to its study and attempted elucidation.

E. H.

[...] [...]

CONTENTS.

PART I.

CHAPTER I.

INTRODUCTION.

CHAPTER II.

PHYSICAL FEATURES.

PART II.

CHAPTER I.

GEOLOGICAL STRUCTURE OF ARABIA PETRÆA AND PALESTINE.

CONTENTS.

CHAPTER II.

ANCIENT CRYSTALLINE ROCKS OF ARABIA PETRÆA.

CHAPTER III.

CARBONIFEROUS BEDS.

CHAPTER IV.

THE CRETACEOUS AND EOCENE BEDS.

CHAPTER V.

MIOCENE BEDS.

CHAPTER VI.

LATER PLIOCENE TO RECENT BEDS (PLUVIAL).

PART III.

CHAPTER I.

TERTIARY VOLCANIC ROCKS.

PART IV.

CHAPTER I.

DYNAMICAL GEOLOGY.

CHAPTER II.

THE PLUVIAL PERIOD (PLIOCENE AND POST-PLIOCENE).

PART V.

CHAPTER I.

ORIGIN OF THE SALTNESS OF THE DEAD SEA.

CHAPTER II.

RECENT CHANGES OF CLIMATE AND THEIR CAUSES.

APPENDIX A.

APPENDIX B.

LIST OF · ILLUSTRATIONS.

PLATES.

WOODCUTS.

PART I.

THE

PHYSICAL GEOLOGY

OF

ARABIA PETRÆA AND PALESTINE.

————◆◦◦◆————

CHAPTER I.

INTRODUCTION.

IT is nearly fifty years since Leopold von Buch expressed the hope that some day the Geological Society of London might send out one of its members to the Dead Sea 'to illuminate with the torch of geology the facts which interest the world.' The work which the illustrious German philosopher destined for a geological society has been undertaken by a much younger, but very active, Society—that of the Palestine Exploration Fund ; and it falls to my lot as the member of the Expedition, sent out by that Society in 1883-84, who had the honour of being entrusted with its direction, to endeavour to lay before its members and the public the conclusions arrived at regarding the physical phenomena which came under our notice during three months' wanderings through a region abounding in matters of interest to the student of nature.

The Expedition, of which a narrative has already been given to the public by the author,* was organized by the Executive Committee of 'The Palestine Exploration Fund,' to make observations along a tract of country extending from the Gulf of Akabah to the Jordan and banks of the Dead Sea, in order to explain the formation of Western Palestine, and of the great line of depression which extends from the base of the Lebanon

---

* 'Mount Seir, Sinai, and Western Palestine' (1885), Bentley and Son.

1—2

on the north to the Gulf of Akabah on the south. The instructions of the Committee were conveyed to the author in the letter of the chairman, Mr. Glaisher, F.R.S., dated 7th July, 1883. This letter prescribed the route to be taken, as far as circumstances would permit, and which was that very nearly carried out in its entirety. We were instructed to proceed by the overland route to Egypt, where we should be joined by Major Kitchener, R.E., and thence to strike into the Desert of Mount Sinai, which we were to traverse as far as the head of the Gulf of Akabah. From thence we were to proceed northwards along the whole length of the Wâdy el Arabah to the southern shore of the Dead Sea, and thence along the western shore as far as Engedi (Ain Jìdi), turn up into the table-land of Judea by Hebron, and thence to Jerusalem. From this central position another expedition through Northern Palestine was to be organized. Circumstances which I have already described obliged us to deviate from the latter part of our prescribed route,* and to abandon that through Northern Palestine altogether; for, just when about to start from Jerusalem on the 21st of January, 1884, a fall of snow took place which covered the whole face of the country to a depth of from one to two feet, rendering travelling almost impossible, and observations on the strata and features of the country either impracticable or deceptive. After waiting for three days, and finding that there was no prospect of the disappearance of the snow for some time, we came to the conclusion that nothing remained but to return home by the first opportunity; and we accordingly embarked at Jaffa on the 26th of January by one of the Austrian Lloyd's steamships, and returned homewards by Beirût, Smyrna, and Constantinople, reaching London on the 13th of February, 1884, the journey having occupied a period of nearly four months.

The members of our party, besides myself, consisted of Major Kitchener, R.E., then on duty in Egypt; my son, E. Gordon Hull, M.D., who was appointed 'Assistant and Hon. Medical Officer;' Mr. John Armstrong (formerly Sergeant-Major, R.E.), who with Major Kitchener had previously been engaged on the Ordnance Survey of Western Palestine; Mr. Henry C. Hart, M.A., who volunteered his services as Naturalist; and Mr. Reginald Laurence, Associate Royal College of

---

* 'Mount Seir,' pp. 118, 127, etc.

Science, Dublin, who also volunteered his services as Meteorologist. My son took charge of the camera for photographs; and, ere leaving London, the members of the Expedition were supplied with all such instruments and apparatus as were likely to be useful for surveying, and making collections of natural history objects. Thus the Expedition was fully equipped; and on Saturday, 10th November, the whole party, with attendants, Arabs, and camels, arrived at the rendezvous of 'Ayun Musa (Moses' Wells) on the borders of the Sinaitic Desert.

Having thus given an account of the objects of the Expedition, and of the members composing it, the reader is referred to the work previously quoted for a narrative of the journey itself.

Works of Previous Authors.—Of the many authors who have, to a greater or less extent, written upon the geology and physical characteristics of the region here described, the following are the more important, or, at least, the more accessible:

ANDERSON, Dr. H. J., 'Geological Report of Palestine' (American Expedition, 1848).

BAUERMAN, H., 'Note on a Geological Reconnaissance made in Arabia Petræa in the Spring of 1868.'—' Quart. Journ. Geol. Soc.,' London, vol. xxv., p. 17.

BURCKHARDT, FRITZ, 'Travels in Syria and Palestine.'

CONDER, CAPT., 'Tent Work in Palestine' and 'Heth and Moab'; also various papers in the 'Quarterly Statement' of the Palestine Exploration Fund.

DAWSON, PROFESSOR SIR J. W., 'Notes on the Geology of the Nile Valley and of Egypt.'—'Geological Magazine,' Nos. 241, 242, 243, 244. New Ser. (1883-84). "Egypt and Syria" (1883).

DAUBENY, PROFESSOR CHARLES, 'Description of Active and Extinct Volcanoes.' 2nd edition (1848), p. 358 *et seq.*

DUNCAN, PROFESSOR MARTIN, 'Fossil Species from the Cretaceous Rocks of Sinai.'— 'Quart. Journ. Geol. Soc.,' London, vol. xxv., p. 44 (1868).

FRAAS, DR. OSCAR, 'Aus dem Orient' ('Geologische beobachtungen').

HITCHCOCK, PROFESSOR E., 'Notes on the Geology of Western Asia, founded on Specimens and Descriptions of American Missionaries.'—'Rep. Assoc. Americ. Geologists' (Boston, 1843).

HAECKEL, PROFESSOR ERNST, 'Arabische Korallen' (1876).

HOLLAND, REV. F. W., 'Journ. Roy. Geog. Soc.' (1866), and 'Notes on the Geology of Sinai.'—'Quart. Journ. Geol. Soc.,' vol. xxii., p. 491 (1866); also, 'Survey of Jerusalem.'

HUDLESTON, W. H., 'Notes on the Geology of Palestine,' etc.—'Proc. Geologists' Association,' Nov., 1882. 'Further Notes on the Geology of Palestine.'—'Nature,' April 30, 1885.

HULL, E., 'Mount Seir, Sinai, and Western Palestine, being a Narrative of a Scientific Expedition,' 1883-84 (London, 1885).

IRBY AND MANGLES, 'Travels in Egypt, Nubia, Syria, and the Holy Land, in 1817-18.'

JAMIESON, T. F., 'Inland Seas and Salt Lakes of the Glacial Period.'—'Geol. Magazine,' May, 1885.

LABORDE, LEON DE, 'Voyage en Orient' (1828); 'Journey through Arabia Petrea,' English translation (1836).

LARTET, LOUIS, 'Voyage d'Exploration à la Mer Morte,' par M. le Duc de Luynes. Tome Troisième.—'Géologie' (Paris, 1880?).

LORTET, L., PROFESSOR, 'Poissons et Reptiles du Lac de Tiberiade,' etc.—'Archives du Muséum d'Histoire Naturelle de Lyon,' t. iii. (1883).

LYNCH, LIEUT., U.S.N., 'Official Report of the U.S. Expedition to Explore the Valley of the Jordan and the Dead Sea' (Baltimore, 1852).

MERRILL, DR. SELAH, "East of the Jordan."

MILNE, JOHN, 'Geological Notes on the Sinaic Peninsula and North-Western Arabia.' —'Quart. Journ. Geol. Soc.,' vol. xxxi., p. 1 (1875).

NIEBUHR, M., 'Beschreibung von Arabien' (1772); 'Travels in Arabia,' 2 vols. (1792).

'Picturesque Palestine,' views of remarkable localities in Palestine, Arabia Petræa, and Egypt (various contributors).

ROBINSON, DR. W., 'Biblical Researches in Palestine,' and 'Physical Geography of the Holy Land' (1865).

ROBERTS, DAVID, 'Illustrations of the Holy Land, Syria, Arabia, Egypt, and Italia' (1842).

RITTER, CARL, 'Vergleichende Erdkunde der Sinai Halbinsel, von Palæstina und Syrien,' 4 vols., 1848-1855.

RUSSEGGER, Joseph, 'Reisen in Europa, Asien, und Afrika,' from 1835 to 1840 (1845 to 1849).—'Journ. N. Leonhardt und Broun.'

SALTER, J. W., 'On a True Coal-plant (Lepidodendron) from Sinai.'—'Quart. Journ. Geol. Soc.,' vol. xxiv., p. 509.

SCHWEINFURTH, DR. VON G., 'Ueber d. Geolg. Schichtengliederung des Mokattam bei Cairo.'—'Zeitsch. d. Deuts. Geolog. Gesellsch. Jahrg.' (1883).

SEETZEN, Ulrich J., 'Brief Account of the Countries adjoining the Lake of Tiberias, the Jordan, and the Dead Sea' (London, 1810).

SCHUBERT, HEINRICH VON (and PROFESSOR ROTH), 'Reise in den Morgenland' (1837).

SMYTH, PROFESSOR WARRINGTON W., 'Presidential Address to the Geological Society of London' (1868).—'Quart. Journ. Geol. Soc.,' vol. xxiv.

TATE, PROFESSOR RALPH, 'On the Age of the Nubian Sandstone.'—'Quart. Journ. Geol. Soc.,' vol. xxvii., p. 404.

TCHIHATCHEF, P. DE, 'On the Deserts of Africa and Asia.'—'Rep. Brit. Assoc.' (1882), p. 356.

TRISTRAM, REV. DR., 'Land of Israel,' 2nd edit. (1872); 'Land of Moab' (1873); 'Fauna and Flora of Palestine' (1884).

WILSON, COL. SIR C. W., 'Ordnance Survey of Sinai' and 'Biblical Gazetteer.'

VOLNEY, Ch., 'Voyage en Syrie et Egypte,' 3rd edit. (Paris, 1880).

ZITTEL, DR. KARL A., 'Ueber den Geologischen Bau der Libyschen Wüste' (Munchen, 1880).

CHAPTER II.

PHYSICAL FEATURES.

THE region embraced within the scope of this essay may be separated into five districts in reference to their physical features, each contrasting strongly with those adjoining, and indicating differences in the geological structure.

1. The first is the maritime district, stretching from the base of Carmel southwards by Ramleh, Gaza, and the Desert of Beersheba, to the Isthmus of Suez, and continued into the plains of Lower Egypt.

2. The second is the table-land of Western Palestine and the Desert of the Tîh ; bounded on the west by the steep descent which leads down into the maritime district of Philistia, and on the east by the limestone escarpments which form the flanks of the Jordan Valley, the Ghor, and the Valley of the Arabah.

3. The third is the depression of the Jordan Valley and of the Ghor,* continued southwards through the Valley of the Arabah into the Gulf of Akabah, attaining its maximum superficial depth below the level of the Mediterranean of 1,292 feet at the margin of the Salt Sea. The name ' Jordan-Arabah Depression ' may be applied to the whole line.

4. The fourth is formed of the elevated plateau of the Jaulan and the tableland of Moab and Edom, which breaks off in a series of abrupt slopes and escarpments along the eastern margin of the Jordan-Arabah depression ; it stretches eastwards into the great Plain of the Euphrates.

5. The fifth district embraces the mountainous tract of the peninsula of Sinai, lying between the Gulfs of Suez and Akabah, and terminating towards the north along the escarpment of the Tîh.

We will now consider each of these physical divisions in the order above stated.

---

* The Ghor (Hollow) is the Arabic name given to the Depression in which lie the Salt Sea and the Jordan south of the Sea of Tiberias.

1. The Maritime District.

This tract consists of a series of low hills from 300 to 400 feet high, separated by valleys and alluvial plains, extending from the Mediterranean sea-board inland to a varying distance. North of Mount Carmel it includes the Plain of Akka (Sanjak Akka), a nearly level expanse, extending from the northern base of Carmel and the banks of the River Kishon (Nahr el Mukutta), by Acre to Ras el Nakurah. The greater part of this tract consists of alluvium and of old sea-beds; and between Acre and the mouth of the Kishon there occurs a line of sand-dunes bordering the coast and piled up by the westerly winds.*

South of Carmel the maritime tract is very narrow for several miles; but at Cæsarea it expands in width, and in the neighbourhood of Ramleh reaches as far as El Kubâb, a distance of about 15 miles from the coast. Southwards, the district still further expands, embracing Philistia and the extensive tracts of sand, loam, and sandstone which slope gradually upwards into the table-land of the Tîh, and terminate towards the west on the shores of Lake Menzaleh.

Throughout the whole of this tract the coast-line is bordered by a line of sand-hills, sometimes rising to a height of 150 feet above the sea; and, where unrestrained by some physical barrier, ever moving inland, being impelled by the prevalent winds from the Mediterranean. The disastrous effects produced by these sand-hills I have already endeavoured to describe;† and Major Kitchener has given a graphic account of their character and effects upon travellers who are obliged to face them when journeying towards Egypt.‡

The maritime district is largely composed of beds of sand and gravel which have once been the bed of the outer sea; while along the line of many of the rivers and streams a deposit of rich loam of a deep brown colour covers considerable areas, and yields luxuriant crops of wheat and maize to the cultivators. The origin of this exceedingly deep loam, which

---

* Shown on the map of the Palestine Survey.

† 'Mount Seir,' p. 145.

‡ Ibid., 'Appendix.' The sands south of Lake Sirbonis, and those extending westwards to the isthmus, are probably in some measure connected with the sands of Lower Egypt and the African Desert.

covers a large tract of country between Beersheba and Gaza, and may be observed in the banks of the Nahr el Hesy, the Nahr Rubln, and the Nahr Nusrah, is not very clear.

The elevation of the maritime district of Philistia varies from 100 to 300 or 400 feet, and at some distance the inland margin may often be very clearly discerned by the uprise of the table-land of Judea. The foundation rock consists of the 'Calcareous Sandstone of Philistia,' which occasionally comes to the surface through the more recent deposits in the form of knolls or low ridges, as is the case at Yazûr, Yebnah, Her-bieh, the hills east of Gaza, and Tel Abu Hareireh. West of the meridian 34° E. long., the margin of the maritime tract is ill-defined, and it gives place to the long terraces and ridges of the limestone plateau.

2. THE TABLE-LAND OF WESTERN PALESTINE AND THE DESERT OF THE TÎH.

This second physical district constitutes the larger portion of the tract under consideration. Bounded on the west by the maritime district, and on the east by the Jordan-Arabah depression, it stretches from the Lebanon on the north to the escarpment of Jebel et Tîh, and Jebel el Ejmeh, along which the limestone breaks off, on the south. It is an elevated plateau, formed almost entirely of beds of limestone, and intersected by numerous ramifying valleys, sometimes narrow and deep like miniature cañons.

The average elevation of the plateau may be taken at 2,000 feet above the Mediterranean; but numerous points and hill-tops reach to much higher levels. Thus, commencing from the north, and proceeding south-wards, we find the following elevations :*

Jebel Hûnî - - -	2,951 English feet.
Safed - - -	2,750 „
Jebel Eslamiyeh, near Shechem -	3,077 „
Tel 'Asûr - - -	3,318 „
Beitîn - - -	2,890 „
Ram Allah - - -	2,879 „
Ras el Mesharif (north of Jerusalem) -	2,685 „
Jerusalem—Temple Area - -	2,593 „

* The elevations are taken from the maps of the Ordnance Survey of Palestine.

Jerusalem—Mount of Olives -	-	2,683	English feet.
Bethlehem -	-	2,550	"
Beit Ummar -	-	3,210	"
Hebron -	-	3,040	"
Beni Naim -	-	3,120	"
Tell et Tûân -	-	2,837	"
Kanân el Aseif (near Tell el Milh)	-	3,002	"
Jebel Magrah (or Mukrâh) -	-	3,460*	"
W. el Hessi (camp)	-	2,170†	"

Watershed.—Throughout the table-land of Western Palestine a line of watershed may be traced from north to south, along which many of the principal towns and villages are planted, and which coincides with one of the ancient highways of the country. Commencing at the ridge which separates the head waters of the Leontes (Nahr el Kasimiyeh) from those of the Jordan, it proceeds along the high rocky ground forming the western edge of the hydrographical basin of the Sea of Tiberias ; it then passes over Mount Tabor (1,799 feet), crosses the head of the Plain of Esdraelon by Mount Gilboa (J. Fukna), where it approaches within six miles of the River Jordan. Thence, turning sharply westward, the watershed passes the head sources of the Nahr Abu Zabura, and proceeds southwards over Mounts Ebal and Gerizim, by Shechem (Nablûs). At Hawâra it bends sharply to east, and running along the ridge bordering the Wâdy Rumeh, again approaches within six miles of the Jordan. Thence, taking a south-westerly course, the watershed may be traced by Bethel to Jerusalem, thence southwards by Bethlehem, the village of El Khudr (2,832 feet), the height called Ras esh Sherifeh (3,228 feet), and Sâfa to Kurbet Es-ha. From this point it takes a sweep towards the east, following the ridge to the east of Wâdy el Khûlil opposite Hebron, and thence stretching south-wards along the ridge of Kanân el Aseif and Khasm Bayûd towards the head of the Wâdy ed Dafâiyeh and onwards towards Jebel Mukrâh in the Badiet et Tîh ; thus separating a system of ancient watercourses which flowed into the Valley of the Jeib from those which drained into the Wâdy el Arish, or 'the River of Egypt,' and through this into the Mediterranean.

The watershed thus described throws off the streams which enter the Mediterranean on the west, and those which descend to the Jordan and

---

* Determined by Mr. Holland.
† Mean of three determinations by Mr. R. Laurence.

Salt Sea on the east. Most of the streams on the west side have a pretty rapid fall near their sources, where they drain the high table-land of Western Palestine; but on reaching the maritime district their waters become sluggish, and they are often bounded by considerable tracts of rich alluvial plain. It is otherwise, however, with the streams which are thrown off on the eastern side, and which debouch on the plain of the Jordan and the Ghor. Owing to the depth of their outlets below the sea-level and the short distance of their sources from the Jordan, the descent of those streams is rapid, and they have often worn down their channels to extraordinary depths. They have been obliged, in fact, to go through a far greater vertical descent than those on the west, and this within a shorter horizontal distance. The impetuosity of these eastern streams at a time when their beds were filled with water must once have been very great, and their eroding effect proportionate. Hence the contrast presented by their channels to those which lie on the western side of the watershed. Many of these eastern streams run for miles between lofty walls of limestone on either side, and ultimately open out on the steep cliffs and escarpments which form the western boundary of the Jordan Valley and of the Ghor. Amongst the more remarkable of these streams are those of the Wâdy el Aujah, which descends from its source at Mezráh esh Sherkiyeh, at about + 3,000 feet to the Jordan at − 1,200 feet, the total fall being 4,200 feet in a distance of 15 miles; or, at the rate of 280 feet per mile. The Kelt (Brook Cherith ?), which, rising at Bireh (Beeroth), at about + 2,800 feet, reaches the Jordan at − 1,170 feet, the total fall being 3,970 feet in a distance of 21 miles, being at the rate of about 190 feet per mile. The stream of the Wâdy el Nar, which, rising east of Jerusalem, at about + 2,400 feet, after passing through the deep gorge of Mar Saba, enters the Salt Sea at − 1,292 feet, the total fall being 3,692 feet in 14 miles; or at the rate of about 264 feet per mile. Besides these are several streams entering the Salt Sea at, or near, Ain Jidi (Engedi), amidst lofty walls of rock, which have a similarly steep descent. Most of these are dry river-beds, except after heavy storms; but, as I shall have occasion to show presently, this was not always the case; for the evidences of great erosive power which these ravines present, bear witness to the existence of a period, now past, when these channels were filled with copious streams, descending in fine cascades and rapids from their sources amongst the Judæan Hills.

3. THE JORDAN-ARABAH DEPRESSION.

The third feature is that remarkable line of depression which, commencing at the north in Cœle-Syria, extends in a southerly direction along the Jordan, the Ghor, and the Wâdy el Arabah into the Gulf of Akabah. That this line of valley coincides with a leading fracture (or 'fault') in the Earth's crust has for some time been recognised ; but we have only here to deal with its outward physical features, and it will be convenient to consider them under two sections, (*a*) the northern, stretching from the sources of the Jordan to the southern end of the Ghor ; and (*b*) the southern, which is continuous from the Ghor to the Ælanitic Gulf.

The extraordinary depression of the Jordan Valley, and of the surface of the Salt Sea below the Mediterranean, remained unproved as far down as the year 1836-7, when Heinrich von Schubert and Professor Roth visited Palestine, and made barometric observations in the Jordan Valley. Thus was demonstrated what had previously been scarcely suspected, that the surface of the Salt Sea is far below that of any other sheet of water on the surface of the globe.*

(*a.*) *The Northern Section.—The Jordan Valley and the Ghor.*

The northern section is drained by the Jordan and its tributaries ; and throughout its length the valley is bounded by abrupt cliffs or terraces on either hand, except where they are cut through by the valleys of the tributary streams.

The Jordan has its source in several springs issuing from the western flank of Jebel es Sheikh (Mount Hermon) ; but the principal fountain is considered to be a small pool a few miles north of Banias (Cæsarea Philippi), in latitude 32° 35′ north, and longitude 33° 26′ east. From these springs the stream descends over a rocky bed to the Lake Huleh, a shallow lagoon the surface of which is 7 feet above the sea-level. Issuing forth from this, it pursues a rapid course of 12 miles along a channel

---

* 'Reise in der Morgenland,' 1836-37.

often bounded by walls of basaltic rock, when it enters the Lake of Tiberias close to the ancient town of Bethsaida. The surface of this lake is 682 feet below the level of the sea, so that there is a total fall of about 689 feet between the two lakes, being at the rate of about 57 feet per mile. On leaving the Lake of Tiberias the Jordan pursues a tortuous course of 66 miles to the Salt Sea, the level of which is (according to the determination of the officers of the Ordnance Survey) 1,292 feet below that of the Mediterranean,* giving a fall of about 610 feet, being at the rate of 9·2 feet per mile. From the Salt Sea it never again issues forth; as all the waters entering from this stream, and those flowing into this great inland lake, pass off into the atmosphere by evaporation. The waters of the Jordan on leaving the Lake of Tiberias are clear; but as the stream washes soft alluvial banks in its course, it soon becomes turgid, as will be observed to be the case at the fords of Jericho.

Owing to the melting of the snows in the Lebanon, the Jordan overflows its banks during February and March.† It is then a broad and deep stream, but during the summer months it is often fordable. Tristram mentions that in January, 1865, the waters of the Jordan had risen 20 feet above the ordinary water-line.‡

The Sea of Tiberias.—This lake, which is nearly oval in form, has a length of about 14 English miles from north to south, and is nearly 8 miles across in its widest part. From its southern extremity the Jordan emerges from amongst lagoons and shallows as a clear stream; the waters of which, as do those of the lake, abound in fishes and molluscs. To the north rise the basaltic hills of Safed; to the east, the volcanic plateau of the Jaulan, dominated by extinct cones and craters, the outlets of the vast lava-streams which have poured themselves down into the Jordan Valley at a very recent period. During the summer most of the streams entering the Lake of Tiberias are nearly dry; but in winter and spring, when they are fed by rains or the melting of the snow, they send down vast quantities of water, which, according to Dr. Lortet, cause the surface of the lake to rise more than 6 feet. The strands are covered by fine gravel formed of little pebbles of limestone, basalt, and rolled flint, polished

---

* Rüssegger made the level 1,341 feet—a very close approximation.
† Josh. iii. 15.   ‡ 'Land of Israel,' 2nd edit., p. 247.

by the incessant movement of the waters, and mixed with innumerable dead shells belonging to the genera *Neritina, Melania, Melanopsis, Cyrena,* and *Unio.* [*]

Terraces.—Throughout its course, from the time the Jordan issues forth from the Lake of Tiberias till it enters the Salt Sea, its channel is cut through alluvial terraces consisting of sand, gravel, and calcareous marl, which sometimes contain shells of the genera *Melania* and *Melanopsis,* in a semi-fossil condition, but of species still living in the Lakes of Tiberias and of Huleh. These terraces are continuous round the shores of the Salt Sea, and form the steep banks through which is cut the gorge of the River Jeib, and by which the depression of the Ghor is terminated along the south. The remarkable terraced hills, formed of saliferous and gypseous materials, called the Lisan and Khasm Usdum, lying on either side of the Salt Sea, are portions of terraces of which the upper surface reaches a level of about 600 feet above that of the Salt Sea. This is the most marked of several terraces formed of ancient alluvial deposits to be found in the Jordan Valley; and points to a period when the waters of the valley stood at a level of over 600 feet above their present ordinary surface.[†] Between the base of the cliffs of Jebel Karantul, near Jericho, and the fords of the Jordan, three of such terraces may be observed, and have already been referred to,[‡]

The first being at a level of 630 to 600 feet.
The second   „   „   520 to 250 „
The third   „   „   200 to 130 „

and below the last-named is the alluvial flat liable to be flooded on the rise of the waters.[§]

The upper surfaces and outer margins of these terraces indicate successive stages, or pauses, in the process of desiccation which the waters of the valley have undergone from the period when they attained their

---

[*] Lortet, *supra cit.,* p. 4.

[†] These terraces are noticed by Lartet, but he under-estimates their levels. 'La Mer Morte,' vol. iii., p. 202.

[‡] 'Mount Seir,' p. 162.

[§] Tristram gives the levels of the terraces in the same district as follow: 1st, 750 feet; 2nd, 550 feet; 3rd, 400 feet; 4th, alluvial flat, 220 feet. All above the Dead Sea. Others are given at Engedi somewhat similar.

maximum elevation to the present day, when they appear to have sunk to their lowest limit. On a former occasion I have shown that originally they reached a level somewhat over that of the Mediterranean, at which time this great inland lake must have stretched from the Lake of Huleh south-wards into the Arabah Valley, and to have had a length of about 200 miles.[*] A terrace formed of gravel, with rolled pebbles, occupies a position to the south-east of Safed. This terrace is at a level as nearly as possible that of the Mediterranean; and, as Dr. Lortet has inferred, indicates that the waters of the Lake of Tiberias formerly stood at the level of those of the outer sea.[†] It would, therefore, correspond to the highest alluvial terraces which occur in the Arabah Valley to the south of the Ghor.

The Salt Sea (or Dead Sea, El Bahr Lut).—This occupies the deepest portion of the great Jordan-Arabah depression, and is enclosed on all sides by terraced hills, except towards the north, where it receives the waters of the Jordan. Several other large streams, both intermittent and perennial, pour their waters directly into this great inland lake, such as the Zerka Maïn (Callirhoë), the Mojib (Arnon), the Hemâd, the Deràah or Kerak, the Kurâhy or Hessi, and others on the east; the Gharundel, the Jeib and the Firkeh on the south; and the Mahanwât, the Seyal, the Areyeh, the Sudeir and others, from the west. The total quantity of water, therefore, entering this reservoir and passing off into the atmosphere by evaporation is enormous; and would be only possible in a region where the air, sweeping over large tracts of waterless and desert land, is deprived of its moisture, and rapidly imbibes that from a large expanse of exposed surface.

The banks of the Salt Sea rise steeply on either hand; those on the Moab side forming a line of terraces composed of sandstone capped by limestone, and deeply furrowed by the brook-courses, which descend from high-level springs of the table-land. The western slopes are composed altogether of beds of limestone with dark chert, except along a portion of the south-western shore, where the terrace of Jebel Usdum, composed of beds of sand, marl and rock-salt, forms an advanced wall along the margin of the waters. The western side, like the eastern, is also deeply channelled by water-courses descending from the table-land of Palestine, generally dry throughout the year.

---

[*] 'Mount Seir,' pp. 99, 108.   [†] Lortet, 'Poissons du Lac de Tiberiade,' p. 4.

The length of the Salt Sea, from north to south, is about 47 English miles, and the breadth, opposite the Lisan, about 10 miles; but the former varies considerably, owing to the existence along the southern margin of a wide expanse of flat shore, generally uncovered by water, but (as shown by the driftwood) liable to be flooded when the waters reach their highest level. According to the observations made on the strand at the foot of Jebel Usdum, the waters rise and fall throughout a height of about 5 feet.

Above the upper limit of the driftwood, and at a height of about 5 feet, there runs a terrace of gravel, extending inland to the cliffs.* This terrace was doubtless under water at a period not very remote, and seems to indicate a continuous desiccation of the atmospheric moisture, and contraction of the waters of the Salt Sea, so that we may well infer they are now at their minimum level.

The great lake is divided into two very unequal portions by a remarkable promontory, which juts out from the eastern shore, and terminates in a cliff facing the west. This is called 'El Lisan,' or 'The Tongue'; and from point Molyneux on the south, to point Costigan on the north, has a length of 9 miles.† The upper surface of the Lisan reaches a level of (according to Tristram‡) about 300 feet. The promontory consists of beds of marl, with thin bands of gypsum and rock-salt, clearly proving that it is a portion of a former bed of the sea, before the falling away of its waters to the present level. M. Lartet, who examined the Lisan carefully, was unable to find any traces of animal remains in the deposits, and draws the conclusion that when the strata were being deposited, the waters were too saline to permit the presence of living organisms.§ The Lisan is connected with the eastern shore by a neck of sand and silt, while the slopes at the base of the Moab escarpment are decked with thickets of small trees and plants, which line the shore to the south of Es Safieh. In the basin to the north of the Lisan, animal life appears to be entirely extinct, and the fishes brought down by the Jordan speedily die on entering the intensely saline waters; but, according to Lartet, in the more shallow and restricted basin to the south of this promontory,

---

* 'Mount Seir,' p. 133.
† In all probability referred to in Josh. xv. 2 as forming the boundary of Judah.
‡ According to Tristram's map.   § 'La Mer Morte,' p. 177 *et seq.*

small fishes belonging to three species of *Cyprinodon* manage to survive,[*] viz. : *C. Moseas*, Cuv. ; *C. Hammonis*, Haeckel ; and *C. Lunatus*, Ehrenb. ; though from Tristram's account it would seem that they only survive in, or close to, the warm streams which flow into the basin.[†]

The surface of the Salt Sea is 1,292 feet below the level of the Mediterranean, as determined by the officers of the Ordnance Survey, and its maximum depth, according to the soundings made by Lieut. Lynch, in 1848, is 1,278 feet, at a point about 5 miles north of Costigan.[‡] The bed was found to descend very rapidly to a depth of 1,050 feet opposite the eastern shore, and thence to vary but slightly till within a mile of the opposite shore towards Engedi. The bottom consists of blue clay, with cubical crystals of salt and lenticular crystals of gypsum ; and it may be inferred that saline and gypseous deposits form the bed to an unknown depth.[§]

The southern boundary of the Ghor is formed by an incurved line of steep banks and cliffs, composed of beds of gravel and sand resting on others of white marl and clay, which are laid open in the banks of the Wâdies Butachy and the Jeib. At the upper edge these banks are about 800 feet above the Salt Sea surface, and from their base extends an extensive plain, descending in low terraces to the great expanse of slimy flats above described as liable to be flooded. The beds forming the banks belong to the extensive lacustrine deposits of the ancient Salt Sea, which gradually ascend to a level corresponding to that of the Mediterranean Sea, as we trace them southwards into the Arabah Valley, and the banks of the 'Ain Abu Werideh.[||]

---

[*] 'La Mer Morte,' p. 270.   [†] 'Fauna and Flora of Palestine,' p. 171.

[‡] The Duc de Luynes found the depth to be 360 metres (1,181 feet) a little to the east of this point, and 300 metres east of Wâdy Mrabbah. 'La Mer Morte,' p. 278.

[§] By an interesting series of observations M. Lartet ascertained that the density and salinity of the waters increase rapidly with the depth—the density at the surface being 1·0216, that at 120 metres being 1·2225, and that at 300 metres being 1·2533.

[||] 'Mount Seir,' p. 99.

(b) The Southern Section.—The Valley of the Arabah.

The Valley of the Arabah, first explored and described by Burckhardt of Basle in 1810, commences where the Ghor terminates—namely, at the edge of the cliffs of sand, gravel, and marl previously described—and extends in a direction south by west to the head of the Gulf of Akabah, a distance of over 100 miles. It presents a marked contrast to the northern section of the Jordan-Arabah depression, owing to the absence of any extensive stream or sheet of water throughout its whole length ; yet many ravines and torrent-valleys debouch on it from either side, and after the heavy storms of winter pour large sheets of water over its surface. Throughout its whole extent the surface of the valley is variously formed of sand, gravel, shingle, and marl, through which the foundation limestone rock occasionally protrudes for short distances. The average breadth of the valley is 6 or 7 miles, but near its centre, to the north of the watershed, it is nearly double this breadth.

The margin of the valley is well defined by the range of Mount Seir on the east, and the escarpment of the Tîh on the west. The eastern range consists of red granite and porphyry, stretching in rugged outline from the banks of the Ælanitic Gulf as far as the Wâdy Tûrban, where the higher elevations are capped by sandstone, which gradually descends in a series of grand terraced escarpments towards the plain. Near the centre of the valley, and on approaching Mount Hor (Jebel Haroun), the sandstone cliffs are several times repeated by faults, and thus lifted to successive levels ; and ultimately the sandstone formation passes below the great terraces of yellow limestone which form the table-land of Edom and Moab, and which rise to a level of about 5,000 feet above the outer sea.*

On the west the Arabah Valley is walled in by beds of limestone, forming a double scarp southwards from the Ghor throughout the greater part of its extent ; but south of the Wâdy Beyâneh the red sandstone formation emerges, and continues to form the lower part of the escarpment of the Tîh as far as the Wâdy el Hendi, where, owing to the effects of faulting, red porphyry, forming a rugged and broken ridge, ranges along the valley-side to the head of the gulf, and onwards to the Râs el Musry.

---

* This is represented in the view of Jebel Nachaleagh and the Edomite Mountains, in Fig. 7, p. 80, of 'Mount Seir.'

A remarkable feature of the Arabah Valley are the extensive fan-shaped deposits of shingle, which have been brought down from the mountain gorges and obtruded on the plain by the torrents. Radiating out from the mouths of the glens, these great sloping sheets of shingle spread themselves over extensive tracts, and being themselves often channelled and furrowed by subsequent torrents, become excessively troublesome to travel over, and cause the camels or horses great fatigue. In other places enormous mounds of pure white sand, rising in dunes 30 to 50 feet high, are met with, not unlike those lining the sea-coast ; and which may be conjectured to have been originally collected at the time when the valley was emerging from the outer and inner seas.

The southern end of the Arabah Valley is formed of beds of marine sand and gravel, containing shells, corals, and other marine forms, stretching upwards in a gentle slope from the shores of the gulf for a distance of about 15 miles. On either side extensive slopes of mountain shingle have been thrown over these more ancient deposits, and to a great extent replace them at the surface. Opposite the 'Ain Gharandel, a deposit of fine silt and sand occupies a depression between two low ridges, the northern of which constitutes the watershed of the Arabah Valley. Amongst the more striking features of this valley is a ridge of por-phyry, known as Samrat el Fiddan, which stretches for several miles in a northerly direction, and terminating in a point of about 1,000 feet above the plain, forms a conspicuous object to travellers coming from the south. The position of the porphyritic ridge is due to the presence of the great ' Jordan Valley fault,' which passes along its western base, and owing to which the limestone strata are to be seen at a much lower level dipping towards the ridge of porphyry. (Fig. 16.) To the west of the ridge, the plain is formed of the old lacustrine strata, which are laid open along the channel of the Jeib. (See Horizontal Section, No. 3.)

The Watershed.—At a distance of about 45 miles from the head of the Gulf of Akabah, the great valley is crossed by a low ridge or watershed, the elevation of which has been determined by several observers.[*] The

---

[*] By M. Vignes of the Duc de Luynes' expedition ; by Colonel Colville, during an expedition in 1883 ; by Mr. R. Laurence, and by Major Kitchener during the expedition of the Palestine Exploration Fund in the same year.

approximate result may be taken at 700 feet, or somewhat less.* Where
the watershed, or rather summit-level of the valley, occurs, a low-scarped
ridge of limestone rises from the plain, and stretching in a north-north-
east direction for a distance of nearly 10 miles, divides the Arabah Valley
into two sections ; the western section of which opens out upon the great
plain of the Tîh, and the eastern forms a depression covered deep by
gravel, and stretching across from the limestone ridge to the base of the
Hills of Edom. Great sand-dunes lie on either side of the ridge, which
thus forms the dividing line between the great depressions of the Jordan
Valley and Salt Sea on the one hand, and of the Ælanitic Gulf on the
other. The water-parting which here crosses the valley has doubtless
continued as such ever since the whole region first emerged from the
ocean ; and is continuous with that which, ranging to the east of the
heights of Edom and Moab into the Syrian Desert, separates the hydrographical basin of the Jordan from that of the Euphrates. Traced in the
opposite direction, this water-parting follows the crest of the inner ridges
which bound the Arabah Valley on the west to the escarpment of Turf
er Rukn,† and continuing along the ridge of Jebel et Tîh and Jebel el
Ejmeh, enters the Sinaic Peninsula at Jebel Dhalal, and the plain lying
between 'Ain el Akhdar and the head of the Wâdy Zelegah. Thence
crossing the ridge by El Watiyeh, it takes a due southerly course through
the mountains to Râs Muhammed, throwing off on either hand the dry
watercourses which drain towards the Gulf of Suez, and those which are
connected with the Gulf of Akabah.

4. THE JAULAN AND THE TABLE-LAND OF MOAB AND EDOM.

In the description of the Jordan-Arabah depression, I have necessarily
touched upon that of the tract now under consideration, which lies along
its margin. Commencing on the north, at the base of Mount Hermon, it
stretches along the borders of the Jordan Valley, through the volcanic
region of the Jaulân, thence into the table-land of Moab and Ammon,
and through Edom, to the mountains of granite and porphyry which rise
to the east of the Arabah Valley and of the Ælanitic Gulf. Throughout
its course of over 300 miles, the tract may be regarded as an elevated

---

* ‘Mount Seir,’ p. 85.   † Ibid., p. 65

table-land, breaking off along the west in a series of slopes and escarpments, which overlook the Jordan-Arabah Valley, and which are deeply
indented on the same side by numerous valleys and ravines, by which
the waters of the springs descend—some to join those of the Jordan and
Salt Sea, others, in the southern portion, to pass off by evaporation into
the air, cr to lose themselves in the sands and gravels which cover the
surface of the Wâdy el Arabah.

Eastwards, this tract merges into the great Syrian and Arabian
Desert. The average elevation of Moab is about 3,000 feet; that of
Edom 4,000 feet, above the Mediterranean. The country is formed of
limestone—the same formation as that of the table-land of Judæa—and
the surface, where not very rocky, is covered by grass south of the Arnon
and west of the meridian of Heshbon; but as we advance eastwards, it
gradually gives way to a low scrub of *Artemisia.* [*]

Amongst the more important heights of this district are Mount Nebo,
near Heshbân, 2,975 feet; Jebel Attarus, over 3,500 feet; Jebel Shihân,
2,780 feet; and Mount Hor (Jebel Haroun), 4,580 feet; at the eastern base
of which lies hidden by surrounding hills the ancient city of Petra, which
gives the name of 'Arabia Petræa' to the whole region.[†] This remarkable city, rediscovered in more recent times by Irby and Mangles, was
thoroughly explored in 1830 by Léon de Laborde, during a period of
eight days, who has left a most interesting account of the ruins, besides
a map of the Wâdy Mûsa, along the course of which the city is built.[‡]
The Wâdy Mûsa was visited by the members of the expedition in 1883,
who had an opportunity of confirming the accuracy of Laborde's map.[§]

The streams which drain the table-lands east of the Jordan-Arabah
depression are remarkable for the evidences they afford of powerful
erosive action. Generally beginning in springs, or in slight depressions
amongst the limestone rocks of the plateau, they deepen their channels as
they approach the western margin; and henceforth the streams bound
onwards in a series of rapids and cascades between lofty walls of rock,
through which their channels have been cut, affording fine sections of

---

[*] W. A. Hayne; quoted by Tristram, 'Land of Moab,' Appendix C.
[†] So named by Strabo, who is followed by Ptolemy.
[‡] A translation of Laborde's work is dated 1836.
[§] 'Mount Seir,' Appendix by Major Kitchener.

the stratification, and enabling the observer to obtain a good insight into the internal structure and constitution of the mountain-side. Amongst the more remarkable of these ravines may be mentioned that of the Zerka, or Jabbok, which drains into the Jordan; that of the Wâdy Heshbân; those of Zerka Ma'in, or Callirrhoë, which enters the Salt Sea;[*] of the Wâdy Mojib, or Arnon; of the Deràah, or Kerak; of the Safieh, or Hessi—all of which enter directly into the Salt Sea; and those of the Wâdy Haroun, Wâdy Mûsa, Wâdies Dalegah and Gharandel, which open out into the Arabah Valley.

Thermal Springs and Streams.—Amongst the most interesting phenomena of the districts just described are the springs of warm water which issue forth on both sides of the Jordan Valley. Of these, the principal are as follows:

THERMAL SPRINGS AND STREAMS.

1. Hammam (or Hammath), near the western side of the Sea of Tiberias. Temperature, 143.3° Fahr.;[†] water sulphurous.

2. Yarmûk Chasm, north of Um Keis, or Gadara. Highest temperature, 109° Fahr.; water sulphurous.[‡]

3. Zerka Ma'in (or Callirrhoë), entering the Salt Sea from the east. Temperature, 130° Fahr.[§]

4. 'Ain Zara,[||] entering east side of Salt Sea. Temperature of water, 109° Fahr. (43° Cent.); temperature of air, 75° Fahr.

5. 'Ain Jidi (Engedi), entering the Salt Sea from the west. Temperature of water, 81° Fahr. (27° Cent.); that of the air, 73° Fahr.

6. 'Ain el Beida; south end of Jebel Usdum. Temperature of water, 91° Fahr.

7. 'Ain W. Khubarah,[¶] entering the Salt Sea from the west. Temperature of water, 88°—93 Fahr.; water sulphurous.

8. 'Ain Feshkhah,[**] west of Salt Sea. Temperature of water, 82° Fahr.

9. Jericho—'Ain es Sultân, west of the Jordan. Temperature of water, 71° Fahr.[††]

---

[*] See excellent views of the deep gorge of this celebrated stream in 'The Land of Moab,' chap. xiii.

[†] Determined by Anderson, of Lieutenant Lynch's Expedition, Off. Rep., p. 202. Tristram, 'Land of Israel,' 2nd edition, p. 432.

[‡] Robinson, 'Phys. Geog. Holy Land,' p. 241.

[§] Tristram, 'Land of Moab,' p. 240. Lartet gives 'Wady Zerka Ma'in, 31° Cent.—88° Fahr.;' but the observations were clearly not taken at the springs.

[||] Lartet, 'Voy. d'Expl.,' p. 291.

[¶] Tristram, 'Land of Israel,' 2nd edit., p. 305.   [**] Ibid., p. 255.

[††] Taken by Mr. R. Laurence, 15th January, 1883, in the pool forming the ancient baths. Doubtless the spring itself would show a higher reading. Others have made the reading 84° Fahr.

From the above selection of the principal thermal springs, it will be observed that those which indicate the highest temperatures are situated along the eastern side of the Jordan Valley. Of these there are five or six, with temperatures varying from 109° to 144° Fahr. ; and it is very significant, as M. Lartet has pointed out, that they are situated along, or close to, the line of the main fault (or fracture) of the valley itself. In the case of the Zerka Ma'in, however, in direct connection with the ancient volcanic rocks, which, 'like those of Staffa,' rise in a series of grand columns above the chasm in which flow the highly heated waters,* it would appear that the subterranean waters are in contact with basaltic rocks, which still retain a portion of their former heat. In a reach of 3 miles there are, according to Tristram, ten principal springs, of which the fifth in descent is the largest; but the seventh and eighth, about half a mile lower down, are the most remarkable, giving forth large supplies of hot sulphurous water. The tenth and last spring is the hottest of all, indicating a temperature of 143° Fahr. Thus it would appear that the heat increases with the depth from the upper surface of the table-land—a result which might be expected, supposing the heated volcanic rocks to be the source of the high temperature itself. To a similar cause may also be attributed the hot springs of Hammath, near Tiberias, and of the Yarmûk (Hieromax) above its confluence with the Jordan.

It can scarcely be doubted, however, that the remaining thermal springs of the east side of the Jordan Valley have their origin in the leading line of fault, which trends along the base of the Moab escarpment. It may be supposed that a large proportion of the rainfall of the table-land, penetrating the porous strata of limestone or sandstone, forms extensive subterranean reservoirs which would have their natural outlet in the Jordan Valley ; but the waters, finding their movement impeded by the walls of the great fissure, burst forth under considerable pressure, and with a temperature due to that of the depth.† That faults in the strata give

---

* Tristram, in his 'Land of Moab,' gives a graphic account of this remarkable gorge and its hot springs, p. 244 *et seq.*

† The rate of increase of temperature due to depth may be taken at 1° Fahr. for every 60 to 70 feet. The increase is estimated from 'The Invariable Stratum' of Humboldt, which approximates to the mean annual temperature of the locality, and at a depth of 50 or 100 feet from the surface.

rise to springs is a fact abundantly verified by numerous observations in several countries.

The warm springs of Jericho ('Ain es Sultân) which issue forth near the foot of the stupendous cliffs of Jebel Karantûl, have their sources in the extensive underground reservoirs of the table-land of Central Judæa. In this case the temperature is probably due solely to the depth from which the waters come below the upper surface. According to Mr. Laurence's determination, the 'Ain es Sultân is 865 feet below the Mediterranean,* and as the average elevation of the table-land to the west is about 2,000 feet above the same datum, the total depth would be about 2,865 feet. Taking the annual mean temperature at 65° Fahr., and adding one degree for each 70 feet, and then deducting 100 feet for the *invariable stratum*, we get a result of 104° Fahr.—that is, 20° higher than that observed.† It is probable, however, the underground stream meets with cooler waters on its passage towards the springs, by which the temperature is reduced. There is evidently no necessity for calling in the aid of volcanic phenomena in order to account for the high temperature ; but it is not improbable the springs burst forth along the lines of fissures parallel to the main fault of the eastern side of the valley.

5. THE FIFTH DISTRICT.—PENINSULA OF SINAI.

This district is remarkably contrasted with that lying to the north ; namely, the table-land of Central Palestine and of the Tîh. Instead of a tract consisting of sheets of limestone terminating in terraced scarps, and traversed by numerous ramifying valleys, the district of Mount Sinai consists of a multitudinous assemblage of rugged heights or sharp ridges, breaking off in cliffs and precipices, and divided by deep waterless ravines ; which, however, form two well-defined systems of drainage, one into the Gulf of Suez, the other into the Gulf of Akabah. The watershed dividing these two systems has already been noticed as taking a due southerly direction from Jebel Dhalal and 'Ain el Akhdar on the north to Ras Muhammed on the south. The district is essentially mountainous,

---

* Taken 14th January, 1884.

† 84° Fahr., according to Baedeker, who, however, does not give his authority. For Mr. Laurence's determination, see p. 22.

rocky, and remarkable for the deep colouration of the granite, porphyry, and schist, of which it is formed. To the north it is bounded by a zone of sandstone, which extends from the base of the limestone escarpment of the Jebels Wutáh, Emreikheh, and El 'Ejmeh, in a band of varying breadth, including the plateau of Debbet er Ramleh, and the banks of the Wâdies Zelegah and Biyar. Towards the west it is bounded by the extensive sandy plain of El Ga'ah, which was formerly the bed of the sea, at a time when the whole region was depressed about 200 feet below its present level.* Between this tract and the sea-coast, the ridge of Jebel Gebeliyeh, formed of Nummulite limestone, stretches for a distance of 35 miles northwards from the Arab village of Tor.

With the exception of Jebel Serbâl, which rises in majestic isolation to the westward, all the principal heights are grouped together near the centre of the peninsula, and a little to the west of the line of watershed. The twin peaks, Katharina-Zebir, are the highest in the peninsula,† the higher rising 8,551 feet above the level of the sea; the other companion heights are Jebel Umm Shomer (8,449 feet), Jebel Mûsa (7,373 feet), and Jebel Umm Alawi, the culminating point of a granite ridge, which, rising abruptly almost as a cliff from the great plain of Es Sened, forms its western boundary.

Separated by about 20 miles of rugged ground from the central group of heights, Jebel Serbâl rises like a rugged wall to a height of 4,000 feet above the valleys at its base, and to that of 6,734 feet above the Gulf of Suez.‡ It is the most striking mountain in the peninsula, and its summit, a ridge of about 3 miles long, is broken into a series of peaks, varying a little in altitude, but rivalling each other in the sharpness and grandeur of their outline.§

Springs.—Though the district of Sinai is considered by geographers as 'rainless,' this is not exactly the case. A little rain falls annually on the highest elevations, and a small sheet of snow rests on the summit of the central group, from which the perennial springs are fed. Six of such

---

* Ordnance Survey Map, 1869.

† Wilson, 'Ordnance Survey of Sinai,' p. 143.

‡ A good representation of Serbâl and the adjoining heights, as seen from the village of Tor, is to be found in Haeckel's 'Arabische Korállen,' Taf. iv., p. 28.

§ Wilson, 'Ordnance Survey, Sinai,' p. 143.

4

rivulets descend from the sides of Jebel Mûsa, while several springs occur
near the higher mountains, as that of Wâdy Aleyât, near the base of Serbâl,
that of Umm Alawi, and that at the head of Wâdy el 'Ain. Running
streams occur in Wâdies Feiran and Sigilliyeh, and for a short distance
in that of El 'Ain.

But the chief sources of supply are the occasional thunderstorms, or
'seils' of the Arabs, which burst on the mountains in December and
January, and give rise to floods which descend the valleys in impetuous
torrents, and with disastrous effect.* That the remarkable gorge of the
Wâdy el 'Ain, traversed by the expedition in 1883, is occasionally the
channel of such floods we had ocular evidence in the large masses of
shingle, boulders, and driftwood piled up at the bendings of the gorge.†
The waters derived from such periodical sources in part sink into the
soil; and, penetrating the rocks, give rise to the occasional springs which
either burst forth naturally, or may be found in digging wells a few feet
below the bed of the valleys.

The rocks of which the Mountains of Sinai are formed are amongst
the most ancient in the world, and may be regarded as the basis upon
which all the newer formations of Arabia Petræa are founded. Their
extremely irregular outline, their deep depressions and lofty heights,
though due in a large degree to atmospheric denudation during geologi-
cally recent times, may have been originally defined during that great
lapse of time between their metamorphism and the Carboniferous period,
when they were largely submerged and covered over by aqueous deposits.
There is every reason to believe that the irregularity of outline is to some
extent continuous beneath the newer formations of the Badiet et Tîh and
of Palestine; and where these ancient rocks are exposed at the base of
the Edomite Mountains, their contour affords evidence of the truth of this
statement.‡

To what extent these mountains, and the neighbouring elevations on
each side of the Red Sea, were enveloped by the foldings of the Cretaceo-

---

* One of these, witnessed by the Rev. F. W. Holland on the 3rd December, 1867, converted the dry bed of the Wâdy Feiran—300 yards wide—into a foaming torrent from 8 to 10
feet deep. 'Ordnance Survey,' p. 243.

† 'Mount Seir,' pp. 58, 59.

‡ See p. 45.

nummulitic rocks, which must originally have rested on their flanks, is uncertain. It seems to me that the higher points must have risen as islands from beneath the waters of the sea at the periods referred to ; and that from the northern base of Serbâl, and from the chain of heights of which this mountain and those of Jebels Mûsa, Umm Alawi, Hammâm and the ridge of Jebel Samghi formed the margin, the great basin in which the Cretaceo-eocene beds were deposited stretched away unbroken for a vast distance towards the north.

Having thus offered a preliminary sketch of the chief physical features of the region embraced by this Memoir, we are in a position to enter upon the consideration of its geological structure, upon which those features so largely depend.

PART II.

CHAPTER I.

GEOLOGICAL STRUCTURE OF ARABIA PETRÆA AND PALESTINE.

IN dealing with the geological formations of which the region under consideration is composed, I propose to describe them in ascending order, beginning with the older, and proceeding on to the more recent, by which means their geological history will be unfolded in the order of events from the remote to those immediately preceding the present age.* The following is the succession of the formations in their descending order :

GEOLOGICAL FORMATIONS.

Recent ....
- 1. Sandhills : Desert sands.
- 2. Alluvial deposits of the Nile and Jordan Valley.
- 3. Gravel of the Arabah Valley.

From Recent to Pliocene.
- 1. Raised beaches : gravel and sand with recent marine shells, etc.
- 2. Ancient deposits of the Salt Sea : marl, sand, and gravel of the Jordan-Arabah Valley.
- 3. Old lake beds : gravel, marl and loam of the valleys of Arabia Petræa.

Miocene ....
- 1. Nicolian sandstone of Jebel el Ahmar, near Cairo.

Eocene ....
- 1. Calcareous sandstone of Philistia, probably of Upper Eocene age.
- 2. Nummulite limestone with chert.

Cretaceous ....
- 1. Cretaceous limestone with chert.
- 2. 'Nubian sandstone' (Neocomien or Cenomanien ?—or Petra sandstone).

Lower Carboniferous ...
- 1. Wâdy Nasb limestone.
- 2. Desert sandstone and conglomerate.

---

* Reference to the geological map is desirable when reading this chapter.

METAMORPHIC ROCKS.

Archæan (?)............... Granite; gneiss; hornblendic, micaceous, and chloritic schists, etc.

VOLCANIC AND PLUTONIC ROCKS.

Recent and Tertiary...... Basalt and dolerite.

Older Primary (?) ...... { Granite, syenite, porphyry, felstone, porphyrite, diorite, basalt, tuff, and agglomerate.

CHAPTER II.

ANCIENT CRYSTALLINE ROCKS OF ARABIA PETRÆA.

OF these rocks, with their accompanying igneous masses, the mountainous region of the Peninsula of Sinai is formed. They also rise into the Mountains of Edom, which stretch along the eastern shore of the Gulf of Akâbah, and the southern flanks of the Arabah Valley; while they are found rising from below the more recent stratified formations along the eastern side of that valley at intervals as far north as the base of Jebel esh Shomrah (or Shomar) at the east side of the Salt Sea. Two isolated tracts, of whose structure and relation to the surrounding Cretaceous beds we have no sufficient knowledge, rise near the centre of the plateau of the Tîh.*

Passing from beyond the limits of our immediate district, it is known that similar rocks form the mountainous tract to the west of the Gulf of Suez and the Red Sea from Jebel Ghareb (Mount Agreb) southwards, and stretch across towards the banks of the Nile to Assouan, crossing that river at the First Cataract, and thence entering the Nubian Desert. Thus these ancient rocks form the general floor upon which have been laid down, at successive ages, all the fossiliferous strata which enter into the structure of Upper and Lower Egypt, Abyssinia, Arabia Petræa and Palestine.

Geological Age.—The crystalline rocks which form the floor of the fossiliferous formations of Arabia Petræa have been referred by Fraas to that ancient period of the earth's history known as Archæan,† and

---

* These are noticed in 'The Field Notes' of the late Rev. F. W. Holland, which have been compiled by Colonel Sir Charles Wilson ('Quarterly Statement Palestine Exploration Fund,' January, 1884, pp. 6, 7). But the references are too meagre to enable us to determine whether they are thrust up along faults, or were old points of elevation in the Cretaceous Sea.

† 'Aus dem Orient,' p. 7.

in Britain and America as Laurentian.[*] As regards the positive evidence afforded in the region of Arabia Petræa itself, it only suffices to prove that these rocks are much more ancient than the Carboniferous, and it might be suggested that they may be metamorphosed Silurian strata. It is exceedingly questionable, however, whether the Silurian basin ever extended over any portion of the Arabian and North-African region. Russegger, Lartet and other observers have recognised in the gneisses, granites, and porphyries of the African Desert bordering the Gulf of Suez, and crossing the Nile at Assouan, the representatives of the crystalline masses of the Sinaitic Peninsula; and Dawson has recently pointed out the remarkable resemblance in mineral character which the former bear to the Laurentian series of North America. Accepting, then, this view, we behold in the Mountains of Sinai a group of rocks of vast antiquity; and which have remained, partially at least, unsubmerged during the enormous lapse of time which intervened between the Laurentian and the Carboniferous epochs. Making every allowance for the effects of denudation, we may with slight hesitation adopt the eloquent language of Fraas, when, pointing to these primæval masses, he exclaims : ' Von Uranfang der Dinge ragten ihre Gipfel aus dem Ocean, unberührt von Silur und Devon, von Dias und Trias, von Jura und Kreide.'

Mineral Characters.—Without a very extended survey, it would be impossible to arrive at a proper knowledge of the stratigraphical succession of these ancient rocks as they occur in the Sinaitic Peninsula. But from the works of previous observers,[†] which my own observations enable me to confirm, it is sufficiently clear that the rocks of which the mountain ranges stretching from the borders of Debbet er Ramleh to that of Ras Muhammed are formed, belong to two great series : the older, metamorphic ; the newer, plutonic or eruptive ; and of these latter, by far the greater part of the mountain ranges appear to be formed.

Metamorphic Series.—These rocks represent the most ancient in Arabia Petræa, and have formed the floor upon which all the other for-

---

[*] Sir J. William Dawson points out that at the First Cataract of the Nile there are two crystalline series, the newer resting unconformably on the older ; the latter he compares with the Laurentian, the former with the 'Arvonian' of Hicks, and the 'Halleflinte' of Sweden.— 'Geological Magazine,' October, 1884.

[†] O. Fraas, p. 7.

mations have been laid down. In coming from Moses' Wells, they are first met with in the Wâdy Nasb, and apparently give place to a grey granite; but their relations to this rock are not very clear. Assuming the granite to underlie the schist, we have the following in descending order :

1. Hornblendic, chloritic, and talcose schists of Wâdies Nasb, Sarabit, Lehean and Feirân.

2. Grey gneiss and granite of Wâdy es Sheikh, and the flanks and summit of Jebel Mûsa.

The schists of the Wâdy Nasb and Wâdy Suwig consist of highly contorted dark greenish beds, penetrated by dykes, and capped by sandstone. They form the lower portion of the east side of the valley, and are bounded by a fault, first noticed by Mr. Bauerman, which runs along the floor of the valley in a direction N. 20° W.[*] In the Wâdy Sarabit, rounded hills of hornblendic schists with mammillated outline are surmounted by sandstone cliffs and terraces. In the Wâdy Kamileh, and as far as the summit-level of Wâdy Bark, hornblendic beds, sometimes exceedingly massive, predominate. These are penetrated by dykes of red granite, or pegmatite, and porphyry, together with a later series of basalt. Indeed, the older and newer masses sometimes become so inter-mixed, that without a detailed examination it would often be impossible to determine which is the more ancient. Near the bend towards the east, in the Wâdy Bark, we come on great masses of red felstone and porphyry, bursting through, and ultimately replacing, the grey fundamental granite or gneiss. In some places this porphyry is columnar, and forms fine cliffs and tors. It consists of a crystalline aggregate of red felspar, blebs of quartz, and green mica, with epidote as an accessory.

Tabular grey granite and gneiss, traversed by thousands of dykes, extends from the grand portals of the Wâdy Berrâh to the ridge of El Watiyeh.

Gneissose rocks form the sides of the Wâdy Aleyat, lying along the

---

[*] Iron and manganese and copper ores were formerly smelted in the Wâdy Nasb and in the Wâdy Baba, and the process of mining is described by Bauerman. Here the mineral occurs under the form of a soft schistose marl variegated with green and red patches, with strings of earthy-brown hæmatite and copper ores.—'Quart. Journ. Geol. Soc.,' vol. xxv. 28.

northern base of Mount Serbâl.[*] These beds, like those of Wâdy Feiran, are penetrated by numerous dykes of granite, porphyry, and diorite.[†] The northern flanks of this magnificent mountain are formed of fine-grained grey gneiss, in which mica forms distinct laminæ. These beds extend upwards to about 3,000 feet below the summit, while above them rises the central mass of beautiful red granite, porphyritic from the presence of large crystals of red orthose felspar, and consisting of orthoclase, oligoclase, quartz, and a little mica, or hornblende.[‡] We may infer that the central mass of granite is eruptive, and belongs to the next group of rocks, presently to be described.

The rocks on which are based the massive cliffs of the Desert Sandstone along the eastern side of the Arabah Valley, are largely formed of various metamorphic schists and varieties of gneiss. This is specially the case between the Wâdies Gharandel and Abu Kuseibeh, near the watershed. At the entrance to the Wâdy Abu Berka the jagged ridges of grey granite, grey gneiss, and mica schist, penetrated by numerous dykes of porphyry, felstone, and melaphyre, are ultimately surmounted with red sandstone and conglomerate. Similar rocks may also be observed east of the plateau of the watershed at the base of Jebel esh Sherah, and the Wâdies Feiran, Solaf, etc.

Plutonic and Volcanic Rocks.—Bursting through the ancient masses of grey granite, gneiss, and schist above described, we find a more recent group of truly igneous rocks, either in the form of dykes, or of masses so large as to form of themselves ridges and mountains. These rocks may be described as porphyries, passing, on the one hand, into granite, and on the other into felstone. They are characterized by depth of colouring, red and purple tints predominating, and giving to the tracts of which they are composed their striking colouration, which the sun's rays have tended to intensify.

Of these rocks are formed the central heights of Serbâl, and of the ranges extending from the ridge of El Watiyeh southwards to Jebel Ghazlani at the southern extremity of the Peninsula of Sinai.[§] They

---

[*] Of which excellent photographic views are shown in ‘ The Ordnance Survey of Sinai.’

[†] Fraas, _supra cit._, p. 11.

[‡] Ibid., pp. 11, 12.

[§] As far as my information, which is very meagre regarding a large portion of the promontory, enables me to judge.

also enter largely into the construction of the ridge of Jebel Samghi bordering the western shore of the Gulf of Akabah, and of the mountains of Arabia on the opposite side of the gulf. They form the boundary ranges of the Arabah Valley from the head of the gulf northwards; on the west side as far as the Wâdy el Hendis, and on the east as far as the Wâdy Dalâgheh, where the granitic rocks give place to others of a fragmental nature, though both probably of volcanic or plutonic origin. The ridge of Samrat el Fiddan is composed of red granite and porphyry, belonging to this group.

Mineral Characters.—This group of rocks is essentially felspathic, felspar of a reddish colour generally forming the paste. It is generally porphyritic from the presence of crystals of felspar, grains of quartz, and crystals of mica or hornblende. Chlorite and epidote are frequently present as accessories; and, under the microscope, small crystals of apatite and magnetite are observable.[*]

Volcanic Fragmental Beds.—In addition to the solid masses above described, which have solidified from a molten state, there are also to be found in several localities volcanic rocks of a fragmental character which are more ancient than the ' Desert Sandstone' of Carboniferous age, and which may be referred with great probability to the period during which the red granites and porphyries were being elaborated in the crust. These latter may be considered as the more deeply seated representatives of the former, which may have been either submarine or subaërial accumulations. They consist of beds of agglomerate, lapilli, and tuff, associated with others of trap.

We first noticed these fragmental masses on ascending the gorge of the Wâdy Haroun, when on our way to visit Petra.[†] There they form the lower portions of the sides of the gorge, and we observed not only the usual dykes of felstone and porphyry traversing the granite, but masses of volcanic agglomerate, consisting of a reddish or purple felspathic paste enclosing blocks and fragments of granite and trap, often angular and of various sizes, such as may have been torn from the sides of a

---

[*] Milne gives a description of specimens from the Wâdy el Ithm and Gebel el Nur; but the above general statement includes the varieties.—'Quart. Journ. Geol. Soc.,' vol. xxxi., p. 19.

[†] 'Mount Seir,' p. 87. Unless the boulder-bed at the entrance to Wâdy Gharandel, described by Dr. E. G. Hull, belongs to the volcanic series.

vent during an eruption. Along with these there are others of fine tuff, and beds of felspathic lava showing an originally viscous, or lamellar, structure.

Jebel esh Shomra'h (or *Shomar*).—A still more interesting exhibition of these ancient volcanic rocks is to be found along the base of Jebel esh Shomrah on the east side of the Ghor, near Es Safieh. The whole base of this mountain, extending for several miles from south to north, is formed of these rocks, which are discordantly overlaid by the Desert sandstone. Here we find an interbedded series of agglomerates and tuffs, together with others of trap—the former fragmental, the latter igneous lavas. The large size of some of the blocks of granite and porphyry, and their rounded form, is remarkable. They seem to have been rolled about under water. The whole of these beds are also traversed by numerous dykes running north and south, which do not apparently enter the overlying sandstone.

The following is a section in part of these beds in the north side of the Wâdy Salmoodh :

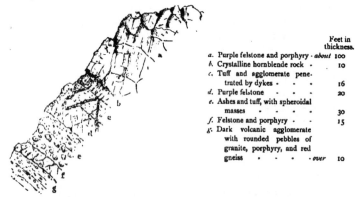

		Feet in thickness.
a.	Purple felstone and porphyry - *about*	100
b.	Crystalline hornblende rock -	10
c.	Tuff and agglomerate penetrated by dykes - - -	16
d.	Purple felstone - - -	20
e.	Ashes and tuff, with spheroidal masses - - - -	30
f.	Felstone and porphyry - -	15
g.	Dark volcanic agglomerate with rounded pebbles of granite, porphyry, and red gneiss - - - - *over*	10

FIG. 1.—SECTION OF ANCIENT VOLCANIC BEDS ON FLANKS OF JEBEL ESH SHOMRAH.

On penetrating for some distance up the deep gorge called the Wâdy Salmoodh, we observed masses of agglomerate and trap so intermixed that we appear to have entered the very throat of the volcano itself. Referring to my note-book I find the following : .

'Where the gorge narrows we pass between lofty cliffs of agglomerate through which numerous dykes of purple felstone and porphyry have penetrated. Some of the blocks (or bombs) are of large size, from 2 to 3 feet in diameter; generally well rounded, and of the most varied composition. We find red, grey, and pink granite; felstone and porphyry of

FIG. 2.—ANCIENT VOLCANIC ROCKS OF WÅDY SALMOODH, EAST SIDE OF THE GHOR.

several varieties, and of red, chocolate, purple, and dark blue colours; fragments of Lydian stone, jasper and schist. The whole mass is confusedly piled together, and penetrated by trap-dykes of short continuance. It seems to me we have here the focus of a volcanic district, and that the fragmental materials are such as may have been torn from the sides of the volcanic throat, while masses of molten lava were forcibly injected amongst these materials from time to time. In several instances the bombs are distinctly embedded in the trap rock.'

The unconformable relations of the Desert Sandstone to the volcanic series can be very clearly made out if the observer takes his stand at some distance from the base of Jebel esh Shomrah near the shore of the Salt Sea. As the flank of the mountain is thus presented, the sandstone beds appear horizontal, and of a bright red or reddish-brown colour, and their base is seen to repose on the dark grey masses of the volcanic beds, penetrated by numerous dykes of trap. The beds of this latter series are clearly bent into a synclinal fold, their denuded edges terminating at the base of the sandstone. The horizontal diagrammatic sketch (Section No. 2) will assist in rendering these relations more clear to the reader.

Trap Dykes.—The whole of these ancient trappean rocks of Arabia Petræa, as well as the still more ancient beds of schist and gneiss, are penetrated by extraordinary numbers of trap dykes, which have called forth remarks from all observers.[*] These are equally remarkable, whether in the Sinaitic Peninsula, or in the hills bordering the Arabah Valley. They consist of two distinct varieties : the older felspathic, and the newer pyroxenic—using the term in its general sense to include hornblendic and augitic varieties. Frequently these two sets of dykes may be observed to intersect each other, in which case the pyroxenic dykes cut across the felspathic. This appears to be an absolute rule without exception, as far as my observation extends ; and they are all more ancient than the ' Desert Sandstone' of Carboniferous age, which they are never seen to enter.

The effect produced by these dykes either when intersecting each other along the sides of a valley or mountain-slope, or when traversing in dark serried lines a tract of country, is most remarkable. Owing to the scantiness of the soil, and of vegetation, their course may often be traced for long distances ; and, as the dykes of basalt or diorite are harder and less destructible than the granitic or schistose masses which they intersect, they often give rise to dark serrated ridges rising along the crest of the hills formed of softer material. One of these remarkable dykes in the Wâdy Lebweh has already been noticed,[†] and a few additional special instances may be mentioned.

Wâdy Kamîleh.—Where the valley becomes narrow the more ancient or fundamental rocks consist of massive hornblende rock, composed of hornblende and white felspar, highly crystalline, passing into syenite by the addition of a little quartz. This is penetrated by dykes of red granite, consisting of quartz, red and white felspar, and black mica ; others of red pegmatite ; these in turn are traversed by a later series of nearly vertical dykes of dark basalt. Near the head ·(or coll) of the valley one of these basaltic dykes is distinctly seen to traverse another of the older granitic series.

Wâdies Berrâh and Es Sheikh.—The granitic walls of the Berrâh valley are remarkable for the numerous basaltic dykes by which they are pene-

---

[*] Fraas, *supra cit.*, pp. 9, 11, etc. Lartet, p. 25. Milne, *supra cit.*, p. 20.
[†] ' Mount Seir,' p. 46.

trated. The granite decomposes into huge oval blocks, with a brownish surface resembling sandstone, and setting off in relief the dark basaltic (or dioritic) dykes. On emerging from this valley into that of Es Sheikh we enter a district of sharp ridges and furrows, of which the fundamental rock is greyish granite, but penetrated in all directions by dykes of red porphyry and felstone, and these by dark basalt or diorite. All these are so inter-laced and numerous that in some places it would be impossible to say which is the predominating rock. The general direction of the basaltic dykes in this part of the Wâdy es Sheikh is west south-west and east north-east, but there are a few transverse to this. Owing to the greater

FIG. 3.—DYKES OF DIORITE OR BASALT TRAVERSING GREY GRANITE IN THE WÂDY ES SHEIKH.

hardness and durability of the dykes than the enclosing granite, they often run along the crests of the little ridges, as represented in the above drawing (Fig. 3). In structure they are often dense and compact, but occasionally porphyritic, and rich in olivine. The granite is also porphyritic, and passes into quartz-porphyry.

Jebel Watiyeh.—But of all the intrusive masses of this region, with the exception of the central ridge of Serbâl, there is surely none more remarkable than the great ridge of Jebel Watiyeh, which, ranging in a direction east by north for many miles, with abrupt flanks and serrated outline, forms, as it were, a breastwork to the mountainous region of Sinai to the south. This ridge, traversed by the remarkable pass of Wâdy el Watiyeh, rises about 1,200 feet above the valley below, and is formed of red porphyritic granite, piercing through the older grey granite or gneiss; as represented in the adjoining sketch (Fig. 4). The junction of the two rocks can be distinctly made out at the northern entrance to the pass; and it is not till we have crossed the ridge, and find ourselves in a district

6

of the older granite again, that the fact of the ridge of El Watiyeh being in reality a huge dyke forces itself upon the mind.

FIG. 4.—GEOLOGICAL SECTION THROUGH JEBEL WATIYEH.

A. Red granite and porphyry breaking through the grey granite.
B. Grey softer granite traversed by dykes of porphyry, D.

The Arabah Valley.—Equally remarkable are the number and varieties of the dykes which penetrate the fundamental granitic rocks along the Arabah Valley, wherever they occupy the sides. I have already given an illustration, taken from the ridge on the west side of the valley, near the head of the Gulf of Akabah,* showing dykes of porphyry and diorite intersecting each other, so as to carve out the granite into lozenge-shaped masses. In a similar way is the red and grey granite of the eastern side intersected by dykes of porphyry, diorite or basalt, parallel to, or crossing, each other, and visible along the naked flanks of the mountains at a distance of several miles. The same is true as regards the granitic base of the ridge of Mount Hor, and the volcanic base of the mountains east of the Ghor, near Safieh, as already described.† Further description would be simple reiteration, and in concluding this part of the subject it may be said that the number and condition of these dykes appear to bear the same relations to the rocks through which they penetrate, that those of Ætna and Vesuvius do to the volcanic masses of those mountains.

---

* 'Mount Seir,' Fig. 6, p. 68.   † *Supra*, p. 38.

Geological Age of the Volcanic Series.—No absolute determination regarding the geological age of the volcanic series is practicable further than this, that they are antecedent to the Lower Carboniferous period. That they are even considerably older may be inferred from the fact that all the crystalline masses above described have undergone an enormous amount of denudation previous to the deposition of the Desert Sandstone. We might go farther, and assume that these plutonic and volcanic rocks are not more recent than the older Palæozoic period, represented by the Silurian rocks of Europe. Whether we may regard them as even more ancient will depend on the date we assign to the older crystalline gneissic and schistose rocks through which they have been intruded. If we consider the former to be of Archæan (or Laurentian) age, then the volcanic series may be of Lower Silurian age ; but if we consider the older schists to be metamorphosed Cambro-Silurian beds, then the newer series may be of Upper Silurian or of Devonian age.[*] On the whole, I incline to the view suggested by Dawson, that the more ancient series are of Archæan, and, inferentially, that the more recent eruptive rocks are of Lower Palæozoic age, or possibly of that known in North America as 'Huronian.'

---

[*] As in the case of those portions of the British Isles where the Lower Silurian rocks have been metamorphosed previous to the deposition of the Upper Silurian beds to which they are unconformable.

The reader is referred to Sir J. W. Dawson's account of the two sets of crystalline rocks, unconformable to each other, at Assouan, which he assigns provisionally to the Laurentian and Huronian stages respectively ('Egypt and Syria,' pp. 22-3). The data for a comparison of the Assouan series with those of Arabia Petræa being insufficient, I am unable, with any degree of certainty, to say whether or not the two series of crystalline rocks in each region are strictly representative of each other; the probabilities, however, lean to this conclusion.

CHAPTER III.

CARBONIFEROUS BEDS.

DESERT SANDSTONE AND WÂDY NASB LIMESTONE.

THE occurrence of a formation of sandstone, capped by beds of limestone containing fossil forms of Lower Carboniferous age, and occupying a position between the ancient crystalline rocks on the one hand, and the Cretaceo-nummulitic formations on the other, has thrown a flood of light on the geological structure of Arabia Petræa. The fossiliferous limestone of the Wâdy Nasb is, as Mr. Bauerman has well expressed it, the key to the geology of the district; and whatever doubt may have existed up to this time regarding the nature of the fossils enclosed in this precious deposit, has been completely set at rest by the determination of the specimens brought home by the Expedition of 1883-84. Had it not been for the presence of this band which lies between two sandstone formations, we should have been unable to recognise with certainty any representative of that vast section of geological time which elapsed between the Archæan, or Huronian, and the commencement of the Cretaceous, epochs. Throughout this enormous interval, which includes the Silurian, Devonian, Carboniferous, Triassic, and Jurassic epochs, the physical history of Arabia Petræa would have been as a blank page to us.

Uneven Floor of the Desert Sandstone.—I may here refer to the evidences afforded that this formation was laid down over an uneven floor of the older rocks upon their partial submergence; and I propose to offer two examples of this; one taken from the Peninsula of Sinai, the other from the Mountains of Edom, east of the Arabah. The illustration (Fig. 5) is taken in the Wâdy Zelegah, about six miles down from its upper end, and one mile below our camp of 23rd November, 1883. Here the beds of brown sandstone are clearly seen near the centre of the valley, terminating bed after bed against the sloping surface of rocks of red porphyry. This section is much more extended than that shown in the figure; and if

FIG. 5.—GEOLOGICAL SECTION IN THE WÂDY ZELEGAH.

viewed from south to north shows the side of the ridge descending under the sandstones.

Another illustration which may be mentioned is one which was visible from our camp of the 12th December, amongst the hills north of Mount Hor (Jebel Haroun). Here the sandstone beds are seen occupying an ancient hollow in the granite and porphyry which forms their general floor, and terminating in cliffs on either hand.

Mineral Characters.—The formation for which I propose the name of 'the Desert Sandstone' consists, as its name implies, of soft sandstone, of purple, red, and variegated colours, often conglomeritic or brecciated.* Its thickness is liable to extreme variation, owing to the fact that it was laid down over an uneven floor formed of the older crystalline rocks, from the waste of which it was itself largely accumulated. Consequently, where the ancient bed was most depressed, the Desert Sandstone occurs in greatest thickness. On the other hand, evidence is afforded that in consequence of ridges or elevations in the original surface of the older rocks, the Desert Sandstone may be sometimes altogether absent, and the overlying Cretaceous sandstone is then the only formation interposed between the crystalline rocks and the limestone of the Cretaceo-Nummulitic age. In the Wâdies Nasb, Sarabit, and Lehean, the thickness of the sandstone varies from 150 to 250 feet.

Wâdy Nasb Limestone.—This interesting stratum occupies the crest

---

* As Mr. Bauerman has shown, the rock in the Wâdy Nasb is much discoloured by the presence of iron, manganese, and copper ores.

of the ridge on the north side of the valley, where first it was discovered by Mr. Bauerman in 1868.* The rock is of a dark-grey colour weathering brown, exceedingly hard and brittle, and about 20 feet in thickness. Fossils (such as corals, crinoids, and brachiopods) are locally abundant, but difficult to dislodge ; and huge blocks have fallen from the crest of the ridge, and cover the flanks. By a fault which runs along the bottom of the valley and close to the spring, the limestone is thrown down on the west side, the displacement being about 400 feet. This fault continues in a north-north-west direction beyond the escarpment of the Tlh, which it intersects at Jebel el Wutâh. The general succession of the strata near our camp in the Wâdy Nasb will be more clearly understood by the following section :

FIG. 6.—GEOLOGICAL SECTION CROSSING WÂDIES SARABIT AND NASB.

1. Nubian Sandstone.
2. Carboniferous Limestone with fossils.
3. Desert Sandstone and Conglomerate (Carboniferous).
4. Metamorphic Schist and Gneiss with dykes of trap.

The hard stratum of limestone may be observed capping the flanks of the valleys leading out of Wâdy Nasb, and sometimes forming terraces at their intersection. In the Wâdy Lehean it is again intersected by a large fault, owing to which the metamorphic schists, penetrated by numerous dykes, are brought into a level with the Desert Sandstone and its cap of limestone on the opposite side. After leaving the Wâdies Lehean and Suwig, we lost sight of the Carboniferous Limestone band, until, to our great surprise, we again discovered it amongst the mountains bordering the Wâdy el Hessi on the east side of the Ghor, interposed between the Desert Sandstone below and the Nubian Sandstone formation above.

Throughout the valleys of El Biyar, El 'Ain, and Et Tihyeh, the sand-

---

* Fossils were shortly afterwards collected from the same beds by the officers of the Ordnance Survey of Sinai.

stone is occasionally seen to rest on uneven surfaces of the older crystalline rocks, and to form tors, or conical hills, sometimes of considerable
elevation—as in the case of Jebel el Aradeh—where the sandstone
surmounts its granitic base, and is itself capped by Cretaceous limestone.
Somewhat similar in conformation are the isolated masses of Jebels el
'Ain and el Berg, which rise about 2,000 feet above the main valley ; but
in these cases the flanks of the mountain are composed of the two sandstone formations, with or without the intervening Wâdy Nasb limestone.*

Jordan-Arabah Valley.—After leaving the Nasb and Suwig Valleys
we were never able to discover traces of the Wâdy Nasb limestone, until
we again came upon it amongst the Hills of Moab east of the Ghor, at
Lebrusch.† For example, it was not discovered between the base of the
Cretaceous rocks east of Turf er Rukn, and the base of the sandstone
series at our camp overlooking the Arabah Valley on the 28th November;
nor, again, between this point and the Cretaceous rocks which we crossed
on descending next day towards Akabah. Neither was it observed
during any of several excursions into the Edomite Mountains—such
as that into the Wâdy Mûsa (Petra)—nor, again, up the Wâdies
Suweireh and El Weibeh.‡ What, therefore, was our surprise to recognise (as we believed at the time) this fossiliferous bed in great force
on ascending to visit the remarkable ruins of Lebrusch, on the banks
of the Hessi.

Here the rock is about 150 feet in thickness, and consists of extremely
hard dark-grey and brown limestone, with fossils in casts. It rests upon
the Desert Sandstone, and is surmounted by beds of the Nubian Sandstone, which forms the upper part of this escarpment, and is, in its turn,
surmounted by the Cretaceous limestones. Our time being limited, we
were unable to collect specimens of the fossils,§ which consisted of crinoid
stems and brachiopods, and could not be dislodged without much difficulty.
The apparently extreme inconstancy of the Carboniferous Limestone may

---

* This limestone is apparently absent along the western escarpment of the Arabah Valley
above the Gulf of Akabah.

† 'Mount Seir,' p. 120.   ‡ Ibid., pp. 101-104.

§ This I greatly regret, as there is, consequently, a slight uncertainty regarding the age
present in my mind.

be regarded as due, partly to erosion previous to the commencement of the Cretaceous epoch as represented by the Nubian Sandstone; and partly to the irregularities in the floor over which the Carboniferous beds were deposited, of which I have already offered some examples.

FIG. 7.—TO ILLUSTRATE THE POSITION OF THE WÅDY NASB LIMESTONE, LEBRUSCH.

A. Nubian Sandstone.
B. Grey fossiliferous Limestone, 150-200 feet thick.
C. Red and variegated Sandstone (Desert Sandstone ?).

From the above remarks, it will be seen that in the absence of the Wâdy Nasb limestone (as far as our observations enable us to judge) amongst the mountains which bound the Arabah Valley along the eastern side, it is impossible to determine with certainty whether any of the red and variegated sandstones which rest on the crystalline rocks of that region are to be referred to the Carboniferous period; the presence of the limestone being the only means by which each of these formations may be certainly recognised.

Fossils.—The fossils collected by the members of the Expedition of 1883, from the limestone of the Wâdy Nasb, have been determined by Professor Sollas, of the University of Dublin, to whom I beg to express my obligations, and from this it will be seen that he agrees with Pro-

---

* It may be presumed that the sandstone below the limestone was the stratum from which the portion of a *Lepidodendron*, 'picked up some years ago by an officer. travelling in Arabia,' and presented to Sir R. I. Murchison, was obtained. The late Mr. Salter named this specimen *L. Mosaicum*, and has described it in the 'Quart. Journ. Geol. Soc.,' vol. xxiv. p. 509 (1868). He considered the specimen as distinctly of Carboniferous age. A *Sigillaria* was also amongst the specimens from the Desert (not Nubian) sandstone.

fessor Tate in his inference regarding the Lower Carboniferous age of this formation. The names are as follows :

ANTHOZOA.—1. *Syringopora ramulosa*, Goldf., occurs in Assize I. of the Carboniferous Limestone of Belgium.

2. *Zaphrentis*, sp. inc.

POLYZOA.—3. *Fenestella* sp. agrees most closely with *F. plebia*, but as the cellules are not preserved it is impossible to be certain of the species (Assize VI., Belgium).

BRACHIOPODA.—4. *Spirifer striatus*, Martin—*S. attenuatus*, Sow. (Assize IV., Belgium).

5. *Productus pustulosus*, Phillips, or *Productus scrabiculus*, Martin. These species are so similar that, without a large series of specimens, it is impossible to distinguish them. Both occur in the same Assize (VI.) in Belgium.

6. *Productus*, sp. inc. It belongs to the '*Striatus*' group of Producti, and may be *P. longispinus*, Phillips.

Professor Sollas writes : ' These determinations leave no doubt as to the Carboniferous age of the ' Desert' sandstone as exposed at Wâdy Nasb, and thus afford a welcome confirmation of the conclusions reached by Mr. Ralph Tate's determination of *Orthis Michelini*, Lév, from the same locality. At the same time Mr. Tate's results show in a striking manner how from a single fossil distinct systems may be sometimes identified. Another point of interest lies in the fact that at Wâdy Nasb we find in the same *gisement* species which in Belgium occur some at the top, some at the bottom, and some in the middle of the Carboniferous limestone. *Orthis Michelini* has a wide range, since it occurs in Assizes I. and IV.; but the species you have brought home are known from single *assizes* only. Is their commingling due to poverty of sediment, so that a few beds in Palestine represent a far larger number in Belgium, or have the fossils in question a wider range than has hitherto been supposed ?'

The following were collected from the limestone of Wâdy Nasb by Captain (now Colonel Sir Charles) Wilson and the late Rev. F. W. Holland, and determined by Professor Ralph Tate :[*]

Orthis Michelini. .	Murchisonia, sp. inc.
Streptorhynchus crenistria.	Eulima (?).
Spirifera (fragments).	Rhodocrinus, sp. inc.
Poteriocrinus, sp. inc.	

---

[*] Ref. *supra*, p. 6.

CHAPTER IV.

THE CRETACEOUS AND EOCENE BEDS.

1. 'THE NUBIAN SANDSTONE.' ('CÉNOMANIEN'? D'ORBIGNY.)

I HAVE adhered to the name, originally given by Russegger to this formation;* it has, however, sometimes been confounded with the older Carboniferous Sandstone just described. Though underlaid by this latter stratum throughout a large portion of Arabia Petræa, in the Valley of the Nile it constitutes the oldest of the more recent formations which rest on the crystalline and metamorphosed rocks of extreme geological antiquity. At Assouan and the First Cataract, in Upper Egypt, the beds of the Nubian Sandstone may be seen reposing directly on the schist, gneiss, granite, and porphyry which are the fundamental rocks of this part of the globe.† But, as Dawson has shown, it is not absolutely certain that part of the sandstone may not be Carboniferous. Between the Nile and the coast of the Red Sea the same formation has been recognised by Fraas, Schweinfurth, and Zittel, forming the flanks of Jebel Hamameh, and stretching northwards to Jebel Tenaseb; but the Lower Carboniferous formation has not as yet been recognised west of the Gulf of Suez.

The designation of 'Monumental Sandstone' (*Grès Monumental*) has been applied by De Rozière to this formation as it occurs in Upper Egypt, where it has been largely employed in the construction of the temples and monuments; but, as Lartet has pointed out, this name is scarcely satisfactory, as the ancient Egyptians employed sandstones of a very different age in the construction of their monuments. On the other hand, the name given by Russegger is sufficiently distinctive, as he has shown that the formation has a wide geographical range in the Nubian Desert. It also stretches eastwards from Sennaar on the Blue Nile. Its

---

* 'Reisen in Europa, Asien und Afrika,' 1845-49.
† Zittel, *supra cit.*, p. 9. Dawson, 'Geol. Mag.,' vol. x., p. 439.

westerly range has been largely extended by Karl Zittel, who, with much probability, assigns it to the Middle or Upper Cretaceous period. No fossils except two species of plants (*Nicolia Ægyptiaca*, Unger, and *Daxodylon Ægyptiacum*, Unger) occur, but in the overlying marls shells of molluscs become plentiful.*

Geographical Range.—In our way from Ayun Mûsa towards Mount Sinai, we first came upon the Nubian Sandstone in the Wâdy Hamr, where, in the sides of the valley, near the eastern base of Sarabut el

FIG. 8.—JUNCTION OF THE CRETACEOUS LIMESTONE AND NUBIAN SANDSTONE—WÂDY HAMR.

L S. Soft yellow Limestone (Cretaceous).
N S. Soft variegated Sandstone (Nubian Sandstone).

Jemel, it may be observed rising at an angle of about 15° from below the calcareous Cretaceous beds (Fig. 8). Here the dip is westward, but on reaching the escarpment of Jebel Wutah the strata swing round and dip at a gentle angle northwards along the escarpment of the Tîh, which is here well defined. Of this formation the plateau of Debbet er Ramleh is mainly composed, and from this it stretches by Jebel Dhalah eastwards, forming the slopes of the Wâdies Zelegah and Biyar.

The Wâdy Zelegah (or *Zelakah*).—This magnificent valley—first traversed by Laborde, afterwards by Palmer, and more recently by the Expedition of 1883—is channelled out chiefly in sandstone and limestone of Cretaceous age, a hard stratum of the latter capping the former, and crowning the escarpments which bound the valley for miles. The sides are often covered by enormous landslips, and masses of rock which

---

* Zittel, *supra cit.*, p. 11. Dawson suspects that the statement of *Nicolia* having been found in this rock in Nubia is an error. 'Geol. Mag.,' No. 243, p. 392.

have fallen from above. The valley is generally perfectly dry, and sometimes is so nearly level that it is with difficulty the eye can detect in which
direction is the fall. There are not wanting, however, evidences that after
one of the occasional thunderstorms which burst upon the plateau of the
Tîh, torrents sweep over the surface with great impetuosity, and the
affluents from the plateau serve to convert a generally waterless valley
into a great river-bed.

Mr. Hart, who climbed the northern side of the valley in a position
about seven miles below our camp of November 22nd, has furnished me
with the measurements (taken by the aneroid) of the several strata of which

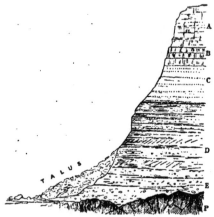

A. Yellow soft limestone.
B. Harder grey limestone, with fossils.
C. Yellowish limestone.
D. Yellow and variegated soft sandstone
E. Red sandstone and conglomerate.
P. Porphyry, etc.

FIG. 9.—SECTION ALONG NORTH SIDE OF THE WÂDY ZELEGAH.

it is formed. The flanks of sandstone are surmounted by beds of limestone,
about 300 feet thick. Where the Zelegah Valley forms a junction with
that of Biyar it becomes an extensive plain, and near the Nawâmîs the
beds are broken off by a large fault ranging east and west, by which the
strata are thrown up on the north side about 500 feet. This fault seems
to have been the immediate cause owing to which the valley changes from
a northerly to an easterly direction.

From thence the Nubian Sandstone may be traced onwards along the
broken tract which leads to the Badiet et Tîh, as far as the escarpmen

of Turf er Rukn,[*] where it forms the flanks and base of the ridge surmounted by beds of limestone and dolomite forming the base of the
Cretaceous series ; and from this it may be traced northwards along the
western flanks of the Arabah Valley as far as Nagb el Salni, where it
disappears under the plain into which the limestone itself descends. On
the east of the Arabah the Nubian Sandstone immediately supports the
Upper Cretaceous strata, forming the table-land of Edom (Mount Seir)
and Moab; but whether it constitutes the entire mass of the sandstone
formation which intervenes between the limestone on the one hand, and
the crystalline rocks on the other, cannot with certainty be determined in
the absence of the distinctive Carboniferous limestone, which was not
observed by us till we reached the banks of the Ghor above Es Safieh.
There can be little doubt, however, that the grand cliffs of variegated
sandstone out of which the ancient city of Petra has been hewn, are
referable in the main to the Nubian Sandstone formation.[†] Along the
eastern slopes of the Ghor and the cliffs bordering the Salt Sea, the
Nubian Sandstone generally constitutes the sides up to a varying level of
several hundred feet from the surface of the lake and bed of the valley,
running far up the lateral ravines ; and it has been traced as far north as
the Wâdy Zerka by Lartet. On the western side of the Jordan Valley and
the Ghor the sandstone never appears, as (owing to the displacement of
the strata by the Jordan-Arabah fault) the limestones of the Cretaceous
period descend to, and below, the plain on that side.

Characters and Composition.—The Nubian Sandstone is remarkable for
the depth and variety of its colouration, due to the presence of various pig-
ments—such as the oxides of iron and manganese, and probably of copper
ores. The uppermost beds, where they emerge from beneath the white marly
limestone in the Wâdy Hamr (*vide supra*, p. 51), consist of dark red, purple,
and brown soft sandstone, stained by oxides of iron and manganese. In the
Zelegah Valley the lower beds are generally white, succeeded by red, and
these by yellowish, strata. I have already attempted to give some idea of
the marvellous colours and patterns which this rock presents in the Wâdy

---

[*] For a view of this feature see ‘Mount Seir,’ p. 65.

[†] By far the best illustrations of these remarkable cliffs are those by David Roberts.
Next to these may be placed the engravings in ‘Picturesque Palestine,’ and the drawings of
Leon de Laborde.

Mûsa, the site of the ancient city of Petra.* Language almost fails
to convey to the reader an idea of the effect produced by the alternations
of yellow, orange, red, and purple tints, of varying depths and arranged in
parallel bands, sometimes straight, at other times waved, or in concentric
curves. Such is the grandeur of form and richness of colouring displayed
by the formation, taken in conjunction with the noble monuments of ancient
art into which the rock has been worked along the sides of the Wâdy Mûsa,
that if one were in want of a name for the formation, none more appropriate
could be suggested than that of ' The Petra Sandstone.'

The base of the formation, where it happens to rest on the older
crystalline rocks, is generally a conglomerate of small pebbles of quartz,
porphyry, or granite ; but the beds which succeed are more or less fine-
grained, and soft. Lartet states that the lower beds contain strata of
shale or marl, but this does not concur with my own observation. The
rock, throughout a thickness of probably not less than 1,000 feet in some
places, seems to be remarkably uniform in composition, and may be
supposed to have resulted from the waste of unsubmerged tracts, composed of crystalline rocks such as those of which the Sinaitic and adjoining
mountain ranges are formed.

2. Cretaceous Limestone (Turonien and Sénonien, D'Orbigny).

The formation of the ' Nubian Sandstone,' just described, indicates the
submergence of extensive areas under the waters of estuaries or restricted
basins ; but that of the calcareous strata (now to be described) points to
a gradually widening and deepening sea-area, together with the submergence of all but the highest elevations of the old crystalline rocks.
The land-areas which existed in various directions during the Cénomanien
epoch gradually receded from view, or disappeared beneath the oceanic
waters ; until at length, towards the close of the Cretaceous epoch, the
waters of the ocean had established their supremacy over a region stretching from the Peninsula of India to, and beyond, the Pillars of Hercules,
and from the northern mountains of Europe and of the British Isles to
the central regions of Africa. The submerged areas included the
northern part of the Desert of Sahara, as far as the Atlas Mountains

---

* ' Mount Seir,' p. 94.

on the west, and the crystalline ranges between the Nile Valley and the Red Sea on the east.* Beyond the Red Sea the Sinaitic and some of the Arabian mountains may also have stood out as land during the Cretaceous epoch, while here and there isolated masses of the central parts of Europe may also have remained as land, at least throughout the greater part of the epoch. The northern part of Arabia Petræa, the whole of Palestine and Syria, together with the regions stretching to the Euphrates, and extending over much of the great Arabian Peninsula, were portions of this vast sea bed, of whose actual limits in several directions we are ignorant, but which we may assume to have been at least as extensive as the basin of the north Atlantic.†

Geographical Range.—In considering the geographical range, it will be convenient to include the overlying Eocene ('Etage Suessonien,' D'Orbigny) limestone beds with the Cretaceous, as, stratigraphically considered, they form but one general formation. The actual area covered by the Cretaceo-nummulitic beds within the region now under description, is sufficiently indicated on the geological map.‡ Its western margin is formed by the wall-like escarpment of the Tîh, stretching from opposite Ismalia southwards to the mountain of Sarabût el Jemel, where the main mass throws off a branch to the western flanks of Serbâl. From Sarabût the margin ranges by Jebel Wûtah in a south-easterly direction to Jebel Dhalal, and from thence in a more or less broken line towards the hills bordering the Arabah Valley. About the head of the gulf of Akabah the limestone forms a capping to detached ridges and outlying hills, about 3,000 feet above the sea-level; but from Ras en Nagb, and the hills above Wâdy Redadi, it forms the crest of the escarpment on the west of the Arabah, and gradually descends till it reaches the plain at Nagb el Salni, south of the watershed. From this point the Cretaceous lime-stones form the western margin of the Jordan-Arabah depression through-out the whole of the distance to the shores of the Sea of Tiberias, where the volcanic rocks descend into the plain.

Within the limits now assigned these limestone beds occupy the region

---

* Zittel, *supra cit.*, p. 9.

† And, of course, twice as large if it included the North Atlantic itself, which is almost certain to have been the case.

‡ Represented by the yellow and buff tints.

of Badiet et Tîh, extending northwards through southern and central Palestine, and through Syria to the Lebanon. They also compose the elevated plateaux of Edom (Mount Seir) and of Moab, surmounting the sandstone of Petra. North of the plain of Jericho they occupy both banks of the Jordan Valley; and on approaching the volcanic districts of the Jaulan and Hauran, the limestone beds are lost to view beneath the vast sheets of basaltic lava which have, at a very recent period, been poured over their surface from the vents of those regions.

Along the Mediterranean sea-board, the Cretaceo-nummulitic beds dip towards the west, and pass below the calcareous sandstone of Philistia from the south of Mount Carmel for an unknown distance beyond Gaza and Beersheba. The western dip is determinable at several points east of Ramleh, and particularly at the Bab el Wâdy. Here the solid beds of limestone (Nos. 2 and 3 of the section below) rise from beneath the upper chalky series. In the district around Beersheba, the strata are but slightly removed from the horizontal position.

Mineral Characters.—Owing to the want of continuous sections, the often disturbed condition of the strata, and the presence of faults and flexures, it is by no means easy to present an accurate account of the general succession of the calcareous beds which enter into the structure of the great table-land extending through Central Palestine to the Badiet et Tîh. Still, from a comparison of various sections, taken at different places and times, the following seems an approximate representation of the series :

GENERAL SUCCESSION OF THE CRETACEO-NUMMULITIC SERIES.

(DESCENDING ORDER.)

Eocene Beds (?) ...	1. White chalky limestones and marls, with occasional bands of dark flint (or chert). Country round Beersheba, Tel el Milh, Sebaste, Nabûlus, etc. (*Nummulites*, bivalves and gasteropods.)
Doubtful	2. Compact grey, yellow, red, and variegated limestone, with marble beds (Jerusalem), with *Ammonites*, *Baculites*, *Turritella* and *Hippurites*.
Cretaceous Beds ...	3. Hard grey, yellowish limestones, sometimes dolomitic, with beds of dark chert or flint, often in considerable quantity. This member is the most important of the Cretaceous series.
	4. Soft white limestone, with rare bands of chert.
	5. Grey calcareous marls passing downwards into shales with selenite and crystals of salt.
	6. 'Nubian sandstone.' Red and variegated sandstone with a base of conglomerate.

And again, in a cliff-section in the flanks of Jebel el Tabrite on the southern borders of the Tîh at Wâdy el Hessi,* where the beds are apparently brought up by faulting. The section is as follows :

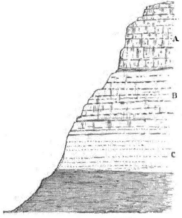

A. Hard silicious limestone with bands
    of chert · · · - 200 feet.
B. Soft white limestone, with rare bands
    of chert · · - 200 „
C. Light blue calcareous marl, passing
    downwards into dark blue clay
    with selenite · · - 250 „

FIG. 10.—SECTION IN CLIFF—WÂDY EL HESSI ; BORDER OF BADIET ET TÎH.

We were unable to discover any fossil-forms in the lower beds, which only were accessible.

The Cretaceous beds may be observed surmounting the sandstone formation along the eastern slopes of the Wâdy Mûsa and the city of Petra, and from thence extending to the summit of the high table-land which reaches an elevation of about 5,000 feet above the level of the sea.† These white calcareous beds may also be observed in a similar position, amongst the deep glens which descend from the table-land of Moab along the eastern banks of the Ghor ; and in these the succession of the strata from the Nubian Sandstone upwards may be advantageously studied.

---

* Examined by our party on the 27th November, 1884.

† It is not improbable that in company with the Nummulitic limestones they underlie the greater part of Arabia—as they have been observed by Dr. Carter on the south-eastern side of the great peninsula forming the ranges of Ras Fartak and Ras Shaawên, bordering the Arabian Sea. The fossils are described by Dr. Martin Duncan.—' Quart. Journ. Geol. Soc.,' vol. xxi., p. 349.

Several of these sections are given in detail by M. Lartet, and amongst
them those of the ravine of 'Ain Mûsa, at the foot of Mount Nebo, and of
the Wâdy Mojib, which, as it commences with the sandstone, I venture
to quote.*

SECTION IN THE WADY MOJIB (LARTET).

(DESCENDING.)

   a. Limestone with red flints at the surface.
   b. Marls with *Pholadomya Luynesi.*
   c. Compact grey limestone.
   d. Alternating chalky marls with *Ammonites Luynesi.*
   e. Friable limestone with *Ostrea Mermeti*, etc.
   f. Grey marls with *Hemiaster Fourneli, Ostrea Olisiponensis*, etc.
   g, i. Calcareous marls and limestones with bivalves and gasteropods.
   j. Limestone with *Ostrea flabellata, O. Africana*, etc.
   k. Green saliferous marls (basal beds).

Nubian Sandstone { *l.* White sandstone.
                 { *m.* Red sandstone.

Cretaceous and Eocene Limestones.—It is admitted by nearly all
observers that the limestones of the Cretaceous and Eocene ages in this
part of the world—when viewed in their mineral aspects—form but one
great series; consequently, without a prolonged and detailed survey, it
would be impossible to separate the two sets of strata, which would often
have to be done on palæontological evidence only.   Oscar Fraas, from an
examination of the fossils of the limestone about Jerusalem, came to the
conclusion that there is a gradual transition from the Cretaceous into the
Tertiary series.   Lartet concurs in this view, and states that while the
presence of *Nummulites* in the calcareous marls (or chalky beds), which
surmount the more solid limestones of the Cretaceous group, shows that
these beds are of Eocene age, their absence, except in special spots,
offers a great difficulty in the attempt to determine the age of certain
beds in particular localities.†   The presence of bands of flint (or chert)
offers no criterion as to age, as they are found both in the Cretaceous
and Eocene strata.‡   In consequence of this uncertainty of identification,

---

* *Supra cit.*, p. 70, Fig. 6, Pl. V. Our party were unable to visit the eastern shores of
the Dead Sea.

† *Supra cit.*, pp. 59, 160, etc.

‡ Lartet states that flint (silex) is found in blocks of Nummulite Limestone both at Mount
Gerizim and at Saida (Sidon), as at other parts of Palestine.

I have represented the margin of the Nummulite limestone on the geological map as only approximate, with the hope that at some future day greater definiteness of detail may be found practicable. Nummulites (*N. Beaumonti*) have been found, according to Lartet, in the limestone of Mount Carmel, in that of Sebastieh (Samaria), and of Nabûlus (Shechem), and from the neighbourhood of Jerusalem.

Dr. Oscar Fraas records the presence of *Nummulites* in the Lower Hippurite limestone of Jerusalem, locally called ' Melekeh,' from the large quarries near the Damascus Gate; from which has been extracted the stone for the ancient walls of the Temple and the Wailing Place of the Jews. The rock is hard, compact, and delicately variegated, and is capable of being cut as marble into objects of ornament and use, and of receiving a polished surface. In his Geological Map of the Environs of Jerusalem (1869), Dr. Fraas gives the following section of the beds below the ' Dépots Quaternaires' in descending order :*

> 1. Craie blanche (*Sénonien*).
> 2. Etage supérieure des Hippurites (' Missih ').
> 3. Etage inférieure des Hippurites (' Melekeh ').
> 4. Zone des *Ammonites rhotamargensis* (Turonien).

The ' Missih,' No. 2 of the above section, affords the principal building stone, and gives the following varieties in descending order :

Broken stuff (Schutt und Trümmer) - - -	5 feet thick.
White stratified marble - - - - -	1 „
Harder stratified marble, from which the frieze of the Sepulchre of the Kings has been hewn - - - -	10 „
Thick oolitic limestone and calcareous marl (' Melekeh ') -	20 „

Colonel Sir C. Wilson has shown that the reservoirs, sepulchres, and cellars under and around Jerusalem are excavated in the softer beds of the ' Melekeh,' which underlie the firmer beds of the ' Missih,' which forms the platform for the buildings.†

That the western portion of the plateau of Badiet et Tîh is mainly formed of Nummulite limestone may be inferred, not only from the observations of the few geologists who have crossed that dreary waste, but on the ground that the limestone series on the east of the Isthmus and Gulf

---

* Aus d. Orient, p. 54.
† ' Ordnance Survey of Jerusalem.'

of Suez represents that of the west, extending from Jebel Attaka to Mokattam south of Cairo, which latter is entirely formed of Eocene strata.*

Nummulites are recorded by Mr. Bauerman in the limestone of Bukel el Faroun,† which rises from the shore of the Gulf of Suez to a height of about 1,500 feet ; and of the Nummulite formation also is formed the tract of Jebel Gebellyeh, which stretches southwards to the village of Tor. Crossing the head of the gulf, we are confronted by the noble cliffs of Jebel Attaka, rising, according to Fraas, about 2,600 feet above the waters of the gulf, and affording a fine section of the stratification. There can be no doubt that in this cliff we have the two formations directly superimposed, and each characterized by sufficiently numerous organisms to ensure identification. The lower portion is opened out in the quarries which were worked by M. de Lesseps for the Suez Canal. The rock consists of greyish-white limestone, with softer beds, most of which are exceedingly rich in fossils, but chiefly in the form of casts. The presence of *Hippurites* serves to determine the Cretaceous age of these beds ; but in those forming the upper part of the cliff, the presence of *Nummulites* and *Cerithium* indicate Tertiary affinities.‡ According to the observations of Sir J. W. Dawson and M. le Vaillant, the strata are traversed by a fault, by which they are repeated in section ; but the main mass consists of Cretaceous beds.

On reaching Jebel Mokattam we find unusual opportunities for the study of nearly the whole series of Eocene strata, which are laid open in the vast quarries from which the building stone of Cairo has been extracted. The observations of several geologists, but more especially of Schweinfurth, have made us familiar with the details of the section, which is composed throughout of Nummulite beds ; so that between the positions of Jebels Attaka and Mokattam the Cretaceous beds gradually descend to a lower level, and on approaching the Valley of the Nile, pass underneath the

---

* Dr. Duncan describes a number of Cretaceous forms from two valleys of the Tîh (Mokatteb and Badera), brought home by the late Rev. F. W. Holland.—' Quart. Journ. Geol. Soc.,' vol. xxiii., p. 38.

† ' Quart. Journ. Geol. Soc.,' vol. xxv., p. 23.

‡ Fraas, *supra cit.*, pp. 110, 111. Fraas gives a graphic sketch of the view over Egypt taken from the summit of this mountain. Dawson concurs in the above identification of the Cretaceous and Eocene beds.—' Geol. Mag.,' No. 243, p. 390.

newer formations and are lost to view, until they emerge (with a northerly dip) from below the Nummulite limestones on the left bank of the river at Thebes and Esneh.*

Paucity of Fossils in the Limestone.—Notwithstanding that the great Cretaceo-nummulitic limestones, like all marine limestone formations, are directly, or remotely, due to organic agency (as Bischof has conclusively shown), it cannot but strike an observer that fossil remains are, on the whole, exceedingly scarce. It is only at intervals that a fossiliferous bed is to be found, and even in this case the forms are sometimes more or less obscure. This arises from the process of transformation which the calcareous matter has undergone; first, during its actual accumulation; and secondly, since its emergence from beneath the waters of the sea. The late Professor Jukes observed this process actually in operation along the borders of the great fringing reef of Australia, where beds of limestone, with but slight traces of organic structure, were being transmuted from the coral rock. Many other examples might be cited in order to show that the preservation of the forms of the shells and skeletons of the marine animals, by whose vital powers the lime has been secreted from the sea-waters, is rather exceptional than otherwise; and we must not, therefore, fly to the conclusion that the absence of such forms is due to any other cause than that of transmutation. Often, however, when no organic structure in a bed of limestone is perceptible to the eye, or under the lens, a thin translucent section placed under the microscope will reveal unexpected forms.†

Flint and Chert.—As Gustav Bischof has shown, all great marine limestone formations have been derived by animal agency from carbonate of lime dissolved in the ocean waters,‡ and as the limestone formations of Judæa and Arabia Petræa are no exception, it follows that the bands and nodules of silicious matter, which are so intimately associated with the calcareous beds, have a somewhat similar origin. These beds of dark chert (or flint) are sometimes specially abundant; and, apparently so, in the

---

* Zittel, ' Geol. Uebersichtskarte d. Libys. u. Arab. Wüste.'

† A thin section of the limestone from the marble quarries of Jerusalem fails to show any well-defined form. It consists of an assemblage of crystals of calcite, which polarize with others of a more granular form; the whole clouded by a little oxide of iron.

‡ 'Elements of Chemical and Physical Geology,' translated by B. H. Paul, vol. iii, pp. 34, 35.

beds which line the banks of the Jordan Valley. Owing to the contrast which their dark tints offer to the whiteness of the associated limestones they provoke observation; and where herbage is scanty amongst the Judæan Hills, and where the strata are at all contorted, they may be observed tracing curves and linear patterns on the hillsides. They are formed of silica, more or less pure, and occasionally contain silicified shells, echinoderms, and even foraminiferæ, which originally were calcareous.[*] Similar silicious bands occur in the limestone formations of other countries. They have their counterparts in the chert beds of the Carboniferous limestone of the British Isles, France, and Belgium; and in the flints in the Chalk formation of Western Europe. Investigations made on the nature and origin of the chert beds of the Carboniferous limestone of Ireland made some years ago by Mr. E. T. Hardman and the author[†] led to the conclusion that their presence was due to the replacement of the original carbonate of lime by silica from a state of solution; similar conclusions have been arrived at by Professor Renard from a study of similar deposits in Belgium. The occurrence of silicified forms of shells, crinoids, echini and corals, which could only have originally been formed of carbonate of lime, proves that they have undergone a change of composition; and as, in general such silicious strata are found in deposits which are eminently free from sandy or clayey sediments, we are forced to refer them to an organic, and subsequent chemical, origin.[‡]

Thickness of the Cretaceo-Nummulitic Limestone Series.—Judging from the large tract overspread by these calcareous deposits, and the numerous flexures and foldings they undergo to the west of the Jordan without revealing their base, as also from the elevation of some of the hills formed of them throughout, it may be inferred that the actual thickness of the limestone series is very great; but there is no means of its accurate determination within our reach. It is probably not an over-estimate if we

---

[*] Drs. Fraas and Roth have observed in silicious beds in the neighbourhood of Jerusalem shells which were certainly originally calcareous, viz.:—*Nautilus zic-zac, Pyramidella canaliculata, Nummulites variolaria* and *N. Biarritzensis.*

[†] Scient. Trans. Roy. Soc., Dublin, 1878. Proc. Roy. Soc. Lond., 1877. Mr. Hardman has noted similar chert beds in the Carboniferous limestone of the Kimberley District, Western Australia.—' Rep. on Geology,' 1884.

[‡] Bischof, quoting the observations of Von der Mark, holds this view as regards the origin of flint in chalk, *supra cit.*, vol. ii., p. 486 *et seq.*

assume a total of 3,000 to 4,000 feet for the Cretaceo-nummulitic series, of which the Nummulite beds are 1,000 feet in thickness.

In the Nubian and Libyan districts of Africa, where the succession of Cretaceo-nummulitic strata can be observed in regular sequence, the thickness has been estimated by Professor Zittel as follows :

				English feet (approximate).
Eocene......	{ Upper Eocene (Schweinfurth) -	-	-	180
	{ Lower Eocene (Zittel)	-	-	1,650
Cretaceous	⎧ Upper chalk (Zittel)	-	-	1,320
	⎪ Greenish laminated clays (Zittel)	-	100 to 264	
	⎨ Beds with *Exogyra Overwegi* (Zittel)	-	-	495
	⎩ Nubian sandstone (Zittel)	-	-	variable

From which it will be seen that the calcareous marine beds attain a thickness of about 3,730 feet along the Nile Valley ; but, as Sir J. W. Dawson has suggested, if the Cretaceous rocks expand in thickness eastwardly, and the Eocene in an opposite direction, it is probable that over the Palestine area the development would be very little different from the above.

3. CALCAREOUS SANDSTONE OF PHILISTIA.

This formation came under my notice for the first time at Tel Abu Hareireh, on which was pitched the camp of our Expedition on the last day of the year 1883, and I have already referred to its discovery.[*] Its presence is the key to the physical features of this part of Palestine, and accounts for the abrupt fall of the table-land of Central Palestine along the borders of Philistia, and along a line extending to the base of Mount Carmel ; as the harder limestones of the table-land dip under, and pass below, the comparatively softer formation of which we are now speaking, and which has been more deeply denuded than the former.

Mineral Characters.—Opportunities were frequently offered of examining this rock between Beersheba and Jaffa ; but, where it is overlain by the terraces of shelly gravel or sand of a later date, and to some extent remodelled, a difficulty was experienced in determining their relations. In general, the strata consist of particles of quartz, cemented by carbonate of lime stained yellow, owing to the presence of oxide of iron. The rock

---

[*] ' Mount Seir,' p. 139.

is sometimes rather solid, but generally porous, distinctly bedded, and uniform in character. Its relations to the more recent beds of shelly gravel at Tel Abu Hareireh are unmistakable, and are represented in the subjoined sketch section, taken on the spot. No fossils were observed in the beds themselves; and at the above place they are seen to be traversed by parallel joint planes, which are absent from the shelly gravels of a more recent period, and indicate their connection with the older calcareous deposits of the country.

FIG. 11.—SECTION TAKEN AT TEL ABU HAREIREH TO SHOW THE RELATIONS OF THE
PHILISTIAN SANDSTONE AND THE MORE MODERN GRAVELS.

S. Calcareous sandstone of Philistia (Upper Eocene).
T. Terrace of shelly gravel and calcareous sandstone (Pliocene to Recent).

Sections of these beds were also observed at ' Sampson's Hill,' near Gaza, and in quarries about three miles north of that town. Here the rock consists of beds of yellowish, calcareous, rather soft sandstone, current-bedded, and containing small fragments of shells. Also at Rhabeh Beit Jerjah, Yebnah, Yazur; and from information furnished by Mr. Armstrong, who was engaged on the Ordnance Survey of this part of Palestine, the boundary between this formation and the underlying limestone has been drawn on the map;* indeed, the physical features themselves indicate this boundary; as it may be assumed to run along the base of the hills constituting the central table-land. The beds are probably in a position but slightly removed from the horizontal throughout the plains of Philistia.

Geological Age.—I have assigned this formation to the Upper Eocene stage, at least provisionally; chiefly on the grounds—(1) that it is older than the sand and gravel of the 200 feet sea-border, which may be inferred to

---

* Thus, from Mr. Armstrong's information I gather that the ridge which runs for several miles from the bridge over the Nahr Rubin is formed of sandstone; also the hill called Dhahr Selmeh, near Jaffa, and the sea-shore cliffs both to the north and south of Cæsarea.

date back to the Pliocene; (2) that there is no evidence of any Miocene beds in Palestine; and (3) that the rock has a very solid character, and is traversed by joint planes, similarly to those of the Cretaceo-nummulitic limestone. The sandstone formation at Tel Abu Hareireh bears a strong outward resemblance to dolomite itself; and it was not till I had broken it with the hammer that I discovered it to be a different rock. The above grounds are admittedly not conclusive; but, from general considerations, may be deemed sufficient, at least for the present.

If we were to search in other districts for representatives, we should probably find them in the Mokattam Hills, near Cairo. There we find a sandstone formation about 180 feet in thickness, forming the uppermost part of the Mokattam beds, and described by Dr. Schweinfurth in terms which are applicable to the sandstone of Philistia. The formation is divided by him into two stages, and of the upper he says: 'Das Gestein dieser obersten Schicht ist ein hellgrauer, bräunlicher oder hellgrauer (*sic*) sandstein, der mehr oder minder kalkreich ist, sich aber stets durch sein fines korn und festes und zugleich poröses Gefüge auszeichnet.'[*]

The underlying member is described in somewhat similar terms, as 'Jener feinkörnige kalksandstein,' in beds of 2 to 3 metres in thickness, and underlaid sometimes by argillaceous beds.—(Thonmergel). The characteristic fossil of this formation is *Echinolampas Crameri;* and the beds themselves are referred by the author to the Upper Eocene stage.

Another sandstone formation, described by Zittel and Schweinfurth, overlies that above described. It is called by the former 'Nicolien Sandstein,' and is considered to be referable to the Miocene stage. It forms a hill called 'Gebel el Ahmar,' lying to the south-east of Cairo, with an elevation of 320 feet above the sea. The sandstone contains the trunks of silicified trees;[†] and although overlying all the Nummulitic beds, is apparently disconnected from them, and passes transgressively over the calcareous sandstone previously described.[‡]

It might be supposed that the calcareous sandstone of Philistia was refer-

---

[*] *Supra cit.*, p. 725.

[†] This interesting locality is also described more recently by Sir J. W. Dawson.—'Geol. Mag.,' Sept., 1884. Also 'Egypt and Syria,' p. 26.

[‡] In some places the Nicolian sandstone rests on stage A A A *a*, and in others on stage A A A 1. See Schweinfurth's Geol. Map of the Mokattam Range.

able to this stage, or at least in part ; and at first I was of this opinion myself : but on further consideration I prefer that already suggested. The sandstone of Gebel el Ahmar is mineralogically unlike that of Philistia. It was probably deposited within the area of an inland lake, while that of Philistia has all the characters of a marine formation deposited in shallow water.

4. LIMESTONE CONGLOMERATE.

On leaving the camp of the Wâdy el Abd on the 28th December, 1883, we visited a large cairn, from which Major Kitchener took a number of angles, on points within sight (see ' Mount Seir,' p. 135). I noticed that the hill-tops about here were overspread by gravel, formed of rounded pebbles of flint, or chert — sometimes several inches in diameter. I was at a loss to account for the rounded and water-worn appearance of these pebbles, till, on finding a section on the hill-side into the rock, I was surprised to find it to consist of soft white limestone enclosing numerous rounded pebbles of flint, evidently water-worn, and forming a veritable conglomerate. As we were now at an elevation of about 3,000 feet above the Salt Sea, it occurred to me that we must be at, or near, the top of the limestone formation ; and that these might be beds formed under the waters of a very shallow sea exposed to wave-action, by which the flint beds had been broken up and rolled about. On the other hand, they might belong to the epoch when the beds were being elevated into dry land. Lartet mentions similar conglomerates occurring on the hill-tops of Palestine.

CHAPTER V.

With the close of the Eocene period, the deposition of strata appears to have ceased over the area now being described, and as M. Lartet well observes, this absence of strata representing the Miocene epoch proves that the region had emerged from beneath the waters of the sea.[*] In accordance with this view, the Miocene period in Palestine and Idumæa was one of disturbance and elevation, of faulting and flexuring, and denudation of strata ; and we can go so far as to say, that it was during this period, which was one of prolonged duration, that the leading physical features of Palestine and Arabia-Petræa began to be developed. Down to the close of the Eocene epoch, the whole region now treated of—together with the mountainous tract of Syria in the north, and a vast extent of Arabia and Northern Africa in the opposite direction—was overspread by the waters of the ocean. With the Miocene epoch land appears ; and as the sea-bed was gradually upraised into the air, and the flexuring and faulting of the strata, due to lateral strain, progressed, the general lines of hill and valley began to be marked out by the help of the various agencies of denudation which simultaneously came into operation.

Miocene Beds of Egypt.—But although absent in Palestine and Arabia Petræa (as far as is known), representatives of the Miocene period have been detected by Zittel,[†] Fraas,[‡] and others,[§] in the Isthmus of Suez, occupying a tract extending east and west for a considerable distance along the base of the Mokattam-Attaka range. They have also been observed in the valley of the Wâdy Ramlieh, south of G. Attaka, where beds of limestone,

---

[*] *Supra cit.*, p. 167. Zittel has indicated the presence of certain Miocene strata in the oasis of Sinah, in the Libyan Desert, and at the foot of the range of hills between Cairo and Suez. They seem to be the last relics of the presence of the sea in those regions.

[†] Zittel, Geol. Map accompanying ' Der. Libysch. Wüste.'

[‡] Fraas, 'Aus dem Orient,' p. 158.   [§] Dawson, 'Geol. Mag.,' No. 243, p. 387.

calcareous marls, sands, and clays, with chalcedony and gypsum, occur, which appear to have been deposited within the waters of an inland lake, after the physical features of the country had been fashioned out of the Cretaceo-Nummulitic beds; and during the epoch of maximum elevation of the Miocene period.   These beds contain fresh-water shells, especially *Ætheria Caillandi,* Férussac, a species now confined to the Upper Nile. In the beds near Gebel Chascah, described by Fraas, bones of cetaceans (like *Halitherium* of the Swiss molasse) and plates of tortoises have been observed.   Along with these are branches and stems of trees, of the family *Asclepias* and broom (ginster).

Of these beds Dawson observes that they would seem to have been deposited 'at a time of Continental elevation, when the isthmus was represented by a wide extent of land, and during the prevalence of a warm climate.'*

---

* It is not impossible that the flint conglomerates which occur on the hill-tops of Palestine may be representative in part of the Eocene epoch.

CHAPTER VI.

LATER PLIOCENE TO RECENT BEDS (PLUVIAL).

THE deposits which now follow for description consist (1) of raised beaches and sea-beds in the maritime districts, and (2) of lacustrine deposits in the interior. It will be convenient to describe them under these two heads :

1. RAISED BEACHES AND SEA-BEDS.

(a) *Maritime Beds.*—Throughout the region bordering the Mediterranean and Red Seas, including the coasts of Asia and Africa, there are to be observed the most clear indications that after the Cretaceo-nummulitic limestone deposits had been raised high into the air during the Miocene epoch, and after the leading features of the land, the main valleys, and coast-lines had been formed by denudation, the whole region was again submerged to a depth of about 220 feet, as compared with the present sea-level.

If we were in imagination to depress this region to the above extent, the waters of the sea would overflow all the plains of Lower Egypt, and ascend up the Nile Valley about as far as the ruins of Luxor and Carnac. They would, also, bathe the terrace on which stand the two great Pyramids, and would flow along the line of the range of hills from Gebel Mokattam to Attaka at the level of the citadel; thence turning along the coasts of the Gulf of Suez, they would overflow the slopes which lead up from the coast to the hills and terraces inland. The great limestone wall which marks the table-land of Badiet et Tîh would once more become the coast-cliffs of the Yam Sûph; and a considerable tract, now occupied by sand-hills along the coast of the desert and as far north as Jaffa, would be submerged. Along this coast the sea would from time to time stretch its arms up the valleys for miles, and form occasional inland lakes connected by narrow channels with the outer sea. To the north of Mount Carmel the Plains of Esdraelon and of Acre would be covered with water, and the limestone ridge on which is built the city of Beyrût would become an island. In a word, the hypothetical problem here presented does actually represent, more or less truly, the relative position of land and sea towards the close of the Pliocene period and onwards. The old sea-beds, now converted into extensive terraces and plains; the old sea-cliffs, now far inland, may everywhere be observed amongst the islands and coasts of the Levant, and along the borders of both arms of the Red Sea.

Let us now proceed to trace the course of this ancient sea-margin, by special cases occurring throughout the district now under consideration from the banks of the Nile eastwards; some from personal observation, the remainder from descriptions of other observers.

(*b*) *Raised Beach of the Nile Valley.*—This remarkable sea-beach is to be observed on both banks of the Nile, where it was first discovered by that indefatigable observer, Dr. Oscar Fraas.* Near the Pyramid of Ghizeh, he observed that the limestone rock, which constitutes the platform on which the pyramid is erected, was pierced by *Pholades* (*Ph. rugosa*, Broc.), and that in a bed of gravel lying against the rock, shells of *Ostrea undata*, Goldf., and *Pecten Dunkeri*, May., were embedded. The position of this bed is about 220 feet above the sea, and corresponds to the level of the sea-margin above Cairo. For many years visitors to the Pyramids had brought away with them specimens of a large sea-urchin (*Clypeaster Ægyptiacus*), but no one knew from whence it was taken. Fraas discovered the spot, and succeeded after three visits in inducing the Arabs to lead him to the habitat of this remarkable fossil. The spot is about two miles southwards from the Sphinx, on the west bank of the Nile Valley, where beds of gravel and marl, containing specimens of this *Clypeaster*, together with shells of oysters (*O. Forskali*) and pectens (*P. benedictus*), may be observed in a horizontal position.† Fraas considered these littoral beds to be of Miocene age; but this opinion is not shared by Schweinfurth or Dawson, who regard them as of later date;

---

* 'Aus dem Orient,' pp. 161-165. It has since been described by Schweinfurth and Dawson; but the latter does not appear to have been aware that the presence of the beach on the west bank is described by Fraas. Sir J. W. Dawson's observations are, however, very acceptable.—'Geol. Mag.,' No. 241, p. 289. Also, 'Egypt and Syria,' p. 26.

† *Supra cit.*, p. 164.

and the presence of *Ostrea Forskali*, and other modern species more recently detected in these beds, appears to bear out this view.

It is clear that the submergence of the Nile Valley, as indicated by these marine shore-beds, must have caused the waters of the sea to ascend far up beyond Cairo; in fact, Sir J. W. Dawson has noticed terraces corresponding to the level of these beaches as far up as Alsilis in the Nile Valley, and he considers the older beds of consolidated gravel at Thebes as belonging to the same Pliocene period.*

(c) *Mokattam Hills, Cairo.*—The next locality is one of extreme interest. It was first recognised by Fraas,† and more recently examined and described by Schweinfurth, who pointed out the traces of the ancient shore-line to the author on the occasion of our visit to Cairo in November, 1883. On ascending the Mokattam Hills towards Gebel el Ahmar, we pass over a tract of undulating ground, and reach the line of the railway from Abbasieh, and here it is that our observations commence. We discover, from certain openings, that the ground is formed of beds of purple and yellow sand and fine gravel, a little marl and clay, with specimens and fragments of *Terebratula* (*T. forscata*), *Ostrea* (*O. cucullata*, Born.), *Pecten*, and *Balanus*—shells or species of which do not occur in the Eocene limestone formation. On crossing the railway and ascending towards the limestone cliffs, we observe that the rock is penetrated by numerous borings of Teredo, though the shell is seldom left in the perforation. We are here evidently standing on the ancient sea-margin, and at an elevation of 220 feet above the Mediterranean and Red Seas. The presence of the beach has been detected in other places along the hills by Dr. Schweinfurth, and the Teredo borings have also been observed by him in the limestone platform on which is built the Mosque of Mehemet Ali.

The level of this sea-beach corresponds pretty nearly with that on the opposite bank of the Nile Valley. It is clear, from its relations to the solid limestone strata, that the Eocene beds had been elevated, and worn back into a line of bold cliffs, marking the limits of the Pliocene Sea in this direction, and that the channel of the Nile Valley itself had been

---

* 'Geol. Mag.,' No. 241, p. 291. These terraces are indicated on Zittel's map; they are outside and beyond the reach of the present river. They seem also to be represented in David Robert's beautiful drawings of the ruins of Thebes and Carnac.

† *Supra cit.*, p. 161.

hollowed out previous to this epoch. And as water finds its own level, we may feel sure that the waters of the Pliocene Sea bathed the flanks of the Egyptian Hills, overspreading the plains of Lower Egypt, and isolating Africa from Asia,* as represented in the sketch map, p. 72.

(*d*) *Suez.*—That the Isthmus of Suez is an old sea-bed has long been known; and was abundantly demonstrated during the progress of the excavations for the Canal. Between the head of the gulf and the Great Bitter Lake beds of recent limestone—somewhat oolitic in texture, and known as 'miliolite'—had to be cut through above the level of high-water mark. These beds contain corals and shells, and amongst them full-sized specimens of *Tridacna gigantea*, with the nacre still fresh. At the landing-stage on the Arabian side, this rock was so hard that it had to be blasted with gunpowder.† Half a mile out of Suez, on the Cairo railway, the raised shell-beach is seen resting on beds of gypseous marl, and rock-salt has been recently found forming an underground stratum of several feet, at a level of 15 feet above the high-water line.‡

The tract of country lying between the limestone escarpment of the Tîh, and the upper part of the Isthmus and Gulf of Suez, is remarkable for the number of terraces which it presents. These may be well observed in the vicinity of Moses' Wells ('Ayûn Mûsa), and sometimes as many as four may be observed in succession. They are formed of sand and limestone gravel, with shells, and their surfaces are often strewn with crystals and plates of selenite. As the escarpment of the Tîh was originally a line of sea-coast, it can scarcely be doubted that these terraces are successive sea-margins. The following shells from the raised sea-bed at Moses' Wells have been kindly named by Mr. Eager S. Smith: *Cerithium erythræonense*, Lamk; *Strombus tricornis*, Lamk; *Nerita crassilabrum*, Smith; *Turbo radiatus*, Grelise; *Trochus erythræus*, Brocchi; *Circe pectinata*, Lamk; *C. Arabica*, Chemnitz; *Chama ruppelii*, Reeve; *Mytilus variabilis*, Krams; *Pectunculus pectiniformis*, Lamk. Also two corals and a Polyzoon.

---

* The large caverns, at a height of about 500 feet above the sea, in Gebel Mokattam, are considered by Dr. Schweinfurth to be connected with a still more ancient sea-margin during the emergence of the land. In this view Dawson concurs.   † Bauerman, *loc. cit.*, p. 18.

‡ Sir John Coode, 'The English Churchman and St. James's Chronicle,' 1st January, 1885. A block of rock-salt, 7 feet thick, was cut from the bottom of the Great Bitter Lake during the excavations for the canal.

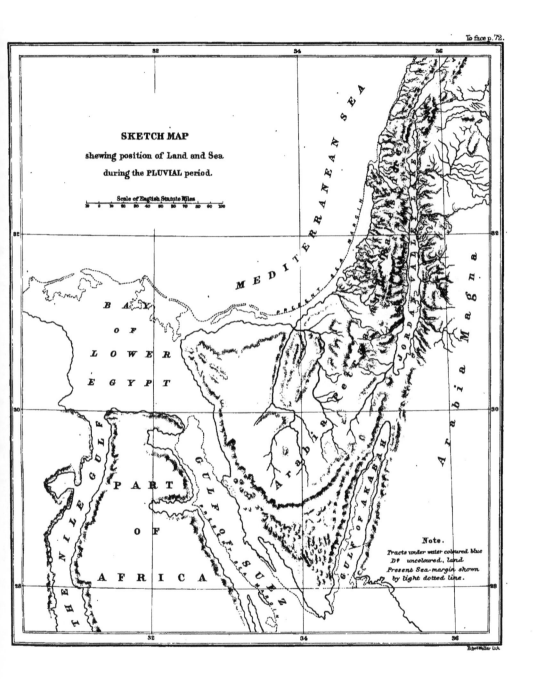

SKETCH MAP

shewing position of Land and Sea

during the PLUVIAL period.

Scale of English Statute Miles

MEDITERRANEAN SEA

BAY

OF

LOWER

EGYPT

PART

OF

AFRICA

THE NILE GULF

GULF OF SUEZ

GULF OF AKABAH

Arabia Petræa

Arabia Magna

JORDAN

PRESENT SEA MARGIN

Note.

Tracts under water coloured blue
D⁰ uncoloured, land
Present Sea-margin shown
by light dotted line.

Edw⁴ Weller lith.

(*e*) The plains of El Makhâ and El Gaâh, which stretch along the eastern shore of the Gulf of Suez to the foot of the interior mountains, are, also, clearly ancient sea-beds.* The latter is a broad and gently undulating plain, covered for the most part by desert sand, while similar tracts of quaternary strata, described on Zittel's map as 'Schuttland und sonstige Alluvialbildungen,' follow the African coast, sometimes running for several miles up the valleys; one of which is the African 'Wâdy el Arabah.'

(*f*) *Akabah.*—As may be inferred, evidences are not wanting that the former sea overspread the lower and southern slopes of the Great Arabah Valley above the head of the Gulf of Akabah. At this locality the ground rises upwards in a gentle slope to the base of the mountains of granite and porphyry on either hand, to a level of about 200 or 250 feet above the sea. These slopes are formed of beds of gravel, amongst which I was so fortunate as to detect specimens of recent coral and sea-shells (amongst them a *Murex*) at a height of about 80 or 90 feet above the sea.† They were obtained from the banks of gravel, where cut down by the little channels formed by the mountain torrents, and were so numerous that I had no doubt that if the sections had been deeper, specimens would have been obtained in abundance. On the day following (Tuesday, 4th December, 1883), Mr. Hart and I found shells near our camp at an elevation of about 130 feet. They included a *Cardium, Trochus*, etc.

FIG. 12.—SECTION TO SHOW POSITION OF RAISED SEA-BED AT AKABAH.

    b, b. Sloping terrace of shelly gravel.
    x, x. Position of shells and coral observed.
    sh. Recent shingle.
    P. Edomite mountains of porphyry and granite.

The relations of these old sea-gravels to the sea on the one hand and the mountains on the other, will be seen from the section (Fig. 12) above. Near the foot of the mountains and the entrance to the valleys,

---

* As shown on the Ordnance Survey Geological Map.
† As I have already stated in 'Mount Seir,' pp. 78, 79.

the marine beds are covered over by great quantities of gravel and detritus brought down from the interior mountains during floods, and spread, fan-like, over the slopes of the valley.

(*g*) *Western Palestine.*—The marine sands and gravels occupy considerable tracts of Western Palestine and Philistia, bordering the sea and running far up the main valleys, which had been channelled out in the older rocks previous to re-submersion during the Pluvial period. The littoral beds rise to levels somewhat over 220 feet above the surface of the Mediterranean; but are largely overspread by the sand-hills which form so remarkable and persistent a feature from the borders of Egypt to the base of Mount Carmel. I shall give some account of special spots.

Wâdy esh Sheríah.—I have already referred to the presence of the marine gravels at Tel Abu Hareireh, in the Valley of Sheríah, near Gaza, and given a section showing their position in reference to the more ancient beds which border the valley. (See Fig. 11, p. 64). Nothing can be clearer than their relations there. The beds occur on both sides of the river valley, in horizontal strata, and at a level of from 200 to 220 feet above the sea. The following is the section on the left bank of the stream opposite our camp of the 31st December, 1883.

SECTION OF RAISED SEA-BED AT TEL ABU HAREIREH, NEAR GAZA.

		ft.	ins.	
1.	Loam - - - - -	5	0	thick.
2.	Soft calcareous sandstone in thin layers - -	10	0	,,
3.	Bed of shells, chiefly in casts - - - -	0	6	,,
4.	Soft calcareous sandstone with small pebbles of flint, and containing small oysters - - - - -	5	0	,,
5.	River-bed—hard calcareous sandstone—thickness unknown (*over*)	2	0	,,

Shells of the following genera were found in the above: *Turritella* (internal spiral cast) *Dentalium, Artemis* (?), *Pecten, Cardium, Ostrea,* spines of *Echinus,* Annelid moulds. We should have much liked to have traced these beds both up and down the river valley, had time permitted.

Gaza.—Around and north of this town by El Mejdel and Esdud, the shelly sands of the old sea-bed may be observed wherever openings occur. They occupy the flats and valleys up to a certain level, and the modern city is itself built over them. Blanched shells of the genera *Cardium* and *Pectunculus* (*P. glycineris*) are the most common forms.

Jaffa.—The raised sea-bed stretches far inland from Jaffa, and is noticed by Lartet.* It may be traced along the Jerusalem road.to beyond Ramleh and Lydda (Ludd). At Jaffa the shelly sands rest on the more ancient sandstone which forms the foundation of the city, and supplies the copious springs of water necessary for the irrigation of the extensive orange and lemon groves which are so justly celebrated for their abundance and excellent quality; but farther inland about Ramleh, this fine sand and gravel gives place to beds of calcareous conglomerate, formed of limestone pebbles of all sizes, and well water-worn. This is undoubtedly an ancient sea-beach, which appears to rise to a level of considerably over 200 feet, formed at a time when the waters of the sea extended over twelve miles inland beyond their present limits. In these beds M. Lartet has noticed the following species : *Pectunculus violascens*, Lamk.; *Purpura hemastoma*, Lamk.; *Murex brandaris*, Linn.; *Columbella rustica*, Lamk., etc. By far the most abundant shell is that first named, and it is still the most abundant on the adjoining Mediterranean shore.

Beyrût.—It has been already stated that the position of the terraces and old sea-beds would lead us to infer that the hill on which Beyrût is built was an island during the Pluvial period.

On crossing the hill, and taking the fine road which connects this city with Damascus, we descend towards a plain which lies between this hill and the advanced spurs of the Lebanon. This plain consists in part of strata which once formed an old sea-beach, but is now a conglomerate, well laid open at a spot by the road-side called Lokandel el Motram.† The conglomerate is formed of water-worn pebbles of limestone from the neighbouring hills, and is cut through by the Beyrût River. It covers an extensive tract in successive terraces, stretching from the base of the Lebanon on the one hand to that of the ridge of Beyrût on the other. These conglomerate beds ascend to an elevation of about 150 feet, and are clearly the floor of an ancient strait which crossed the neck of land, and connected the waters of the bay to the north with those of the open sea.

(*h*) *Other Localities.*—Though beyond the limits assigned to this work, it may be well to mention two additional localities where old sea-beds referable to recent times have been observed, as indicating the

---

* *In loc. cit.,* p. 170.
† Where it was shown to me by Rev. Dr. Bliss, Principal of the American College.

very general depression of the area now bathed by the waters of the Levant. One of these occurs amongst the valleys of the Lebanon in the vicinity of Lattakia (Ladikeyeh), described in detail by Dr. Post, of Beyrût.* According to this account beds of sea-shells and corals now living in the Mediterranean occur at elevations chiefly between 150 and 250 feet above the sea; but they are also found more sparingly at even higher levels. The principal locality where these shells have been observed is in a valley near the village of Qutrûjeh, in a mass of unsolidified clay, which is extremely full both of shells and corals.

Those who have visited the Isle of Cyprus will have noticed the broad terrace which stretches along Larnika Bay, bounded inland by a line of white limestone cliffs. This terrace is also an old sea-bed, and the cliffs formed the coast-line which were washed by the waves at a time when the land was submerged. These raised sea-beds have been described by Mr. R. Russell, who recognises in them shells of species now living in the adjoining waters of the Mediterranean. Older still than these are certain beds of sand, calcareous sandstone and conglomerate, which he refers to the Pliocene period, and which are unconformable to still more ancient chalky strata of uncertain, but possibly, Miocene age.†

Thus it will be seen that all along the coast of the Levant, from Egypt, by Palestine, Syria, and extending to the Island of Cyprus, there are indications that at a period, so recent that the shells and corals are still living, the land has been submerged to a depth of from 220 to 250 feet. During this period Africa was an island, and the waters of the Mediterranean stretched southwards into the Red Sea. This brings us to the consideration of a very interesting problem regarding the cause of the striking dissimilarity between the faunas and floras of these two seas at the present day.

(i) *Cause of difference between the Faunas of the Red Sea and of the Mediterranean.*‡—As Professor Haeckel has observed, the fauna and flora of the Red Sea and of the Mediterranean have developed quite independently of each other; those of the Red Sea belong to the Indian Ocean, while those of the Mediterranean are representative of the

---

* 'Nature,' August 21, 1884.

† Trans. British Assoc., 1881.

‡ I have examined this point at some length in 'Nature,' April 30, 1885.

Atlantic.* Yet the question arises, if these two seas had so recently been united, how are we to account for the great dissimilarity in the forms which inhabit their waters respectively ? The proportion of species of molluscs common to the two seas is very small, being less than 2 per cent.† It might have been supposed that their recent connection through the Straits of Suez would have resulted in a much larger number of common species. If we carefully trace the life-history of the animals inhabiting these two great seas, we shall probably find that the problem is capable of a satisfactory solution ; and it is scarcely necessary to observe that it is only on strictly physical and biological principles that the problem can be solved.

Down to the close of the Eocene period, the whole region bordering the Mediterranean, and including large portions of the three Continents, together with the sea itself, were covered by the waters of the ocean, and a community of species existed, the representatives being now preserved in a fossil state in the strata of the Nummulite limestone series. But with the succeeding Miocene period a change took place over the whole area. The sea-bed was largely elevated into dry land ; while, at the same time,

---

* 'Arabische Korallen,' p. 42. Haeckel observes : ' Die Fische, Krebse, Mollusken, Sternthiere, Würmer und Pflanzenthiere des rothen Meeres zum grössten theile völlig ver-schieden von denen des nahen Mittelmeeres.' Compare Klunzinger, ' Über die Fauna des rothen Meeres,' in Verhand. d. zoolog.-botanisch. Gesellschaft in Wien.

† Professor Issel, of Genoa, in his ' Malacologia del Mar Rosso,' 1869, concludes that there are 18 species of molluscs common to the Red Sea and Mediterranean. He rejects the larger number, 74, given in Woodward's ' Manual' from the collections of Hemprich and Ehrenberg, on the ground of the mixed condition of their collections in the Berlin Museum, the labels being sometimes simply ' Egypt' without defining whether northern coast or Red Sea. Issel also enumerates 30 species of each province which he regards as representative, or only differing so slightly that they may be looked upon as recently modified on one side or the other. His total of the Red Sea is 640 species, and from the above and further evidence which he adduces, he concludes that during the Pliocene and Miocene periods the Mediterranean received some portions of the Red Sea fauna, and was connected with it.

In 1870, Robert MacAndrew published in the ' Annals and Magazine of Natural History,' a report on the testaceous molluscs collected by him in the Gulf of Suez. His total of 995 named species adds 355 to Issel's total. Further, if we accept 199 undetermined species, mostly probably new, and all, as he states, additions to the Red Sea fauna, we have a grand total of (640 + 355 + 199 =) 1,194 recent molluscous inhabitants of the Red Sea. Moreover, from the number of Mediterranean species in MacAndrew's list, the slight community of species regarded as a percentage of the whole Red Sea molluscous fauna is reduced from nearly 3 to less than 2 per cent.—Note by Mr. H. C. Hart.

other portions became deeper, and the general outlines of the present lands and seas were marked out. It was at this epoch that the two great seas became disconnected, and henceforth the fauna of both seas developed independently of each other. Throughout the whole of the Miocene period, and a portion of the succeeding Pliocene, the differentiation of the respective faunas proceeded quite independently of each other, and possibly to the extent of altering a very large proportion of the species. The Miocene period was one of vast duration; and once the Mediterranean and Red Seas became disconnected, the condition of their waters would tend to vary from each other, that of the former becoming colder and that of the latter becoming warmer as time went on.*

It may, therefore, be inferred that at the time of the Pliocene and Pluvial epochs, a large dissimilarity between the species of the existing seas had been established through natural processes; and, when again the isthmus was submerged to the extent, as above shown, of about 220 to 250 feet, this amount of subsidence would have been altogether insufficient to bring about a general commingling of forms. The ridges of land extending from Tel el Kebir and across by El Guisr and Tunum north of Ismalia, and now rising about 50 feet above the sea, would have still further diminished the depth of water stretching over the isthmus.† We may, therefore, assume that during the period of greatest submergence the connecting strait was not deeper than 150 to 170 feet at its shallowest part; and it is scarcely necessary to observe that so shallow a channel would have been quite insufficient to allow of a large interchange of the faunas of the two seas. Only the littoral forms could have crossed from sea to sea in an adult state; all the medial and deep sea

---

* The difference of temperature in the waters of the two seas must have become considerable even in Miocene times. The Mediterranean waters would receive streams draining the continental lands lying to the north of its borders—of a comparatively low temperature—while those entering the Red Sea would draw their supplies from the subtropical districts of Africa and Asia, with a considerably higher temperature. Added to this would be the effects due to difference of latitude in each case. The deeper waters of the Mediterranean at the present day are known, from the observations of Dr. Carpenter, to remain at a comparatively low temperature throughout the year, namely, a little over or under 12° C (54·7° Fahr.).—Wyville Thomson's 'Depths of the Sea,' p. 190.

† This is supposing the present levels were approximately the same as those of the period of submergence—a supposition which, though not capable of proof, may be admitted as approximately correct. During the 500 feet submergence the depth of water over the isthmus would have been proportionately greater. (See note, p. 71.)

forms would have been effectually barred off on either side.[*] The process of differentiation could, therefore, have proceeded during the Pliocene and Pluvial periods, almost as completely as during the Miocene period, and have gone on uninterruptedly down to the present day. Hence the remarkable dissimilarity of the two faunas ; so great, in fact, that were the beds of the seas themselves elevated into land, and their contents presented for our inspection in a fossil state, we would certainly conclude they were representative of two entirely different epochs.[†]

Here, then, we have the case of two large, and almost entirely distinct, faunas developing from one original stock, within a period between the close of the Eocene and the commencement of the present epochs.[‡]

2. LACUSTRINE BEDS.

Inland Representatives of the Pluvial Period.—The inland representatives of the Pluvial period consist of old lake-terraces, whether along the line of the Jordan-Arabah Valley, or amongst the valleys of the Sinaitic Peninsula. We shall consider them in the order here named.

(a) *Old Lake-terraces of the Jordan-Arabah Valley.*—I have already, to a certain extent, described these deposits formed over the bed of the great Jordan Valley Lake at a time when its waters stood somewhat higher than the level of the surface of the Mediterranean, and at successive periods when its waters were gradually falling to lower and lower levels during the process of desiccation and contraction, which has now apparently reached its maximum.

In a former volume,[§] I have attempted to describe the interest with

---

[*] In connection with this point it is interesting to observe that the sub-fossil forms embedded in the strata under the Bitter Lakes, and the raised sea-beds of the isthmus south of the ridge of Tunum, are representatives of those of the Red Sea, while those north of Tunum are representatives of the Mediterranean. The ridge of Tunum seems to have been an effectual barrier to the fauna of the two seas during submergence.

[†] It has been suggested to me by Professor Hadden that many of these deeper-sea forms would have crossed the straits in the young or larval state ; but if so, the difference in the temperature of the two seas would have proved inimical to their existence.

[‡] If the reader will compare the list of Mediterranean mollusca by Mr. J. Gwyn Jeffreys ('Annals and Magazine of Natural History,' vol. vi., p. 65, 1870), with that of the Gulf of Suez by the late Mr. Robert MacAndrew (ibid., p. 428), or with the same as republished by Mr. A. H. Cooke (ibid., vol. xv., p. 322, 1885), he will become fully aware of the extreme specific dissimilarity between the mollusca of the two seas, which also extends to other forms of marine life.   [§] 'Mount Seir,' p. 99.

which on reaching our camp of 'Ain Abu Werideh on the 12th December, 1883, we first observed the horizontal terraces of marl, silt, and gravel with which we were on all sides confronted. The aneroid indicated that we were nearly on a level with the outer sea (rather above than below), or about 1,400 feet above the surface of the Salt Sea, towards which we were marching ;[*] yet as there was no ridge or barrier between us and the shore of that great inland lake, it became absolutely certain that these were lacustrine beds, which had been deposited at a time when its waters stood about 1,400 feet above their present surface.

FIG. 13.—TERRACES OF MARL OF THE ANCIENT JORDAN-ARABAH LAKE-BED AT 'AIN ABU WERIDEH.
Terraces of marl with shells, 20 feet high.

The terraces at 'Ain Abu Werideh are laid open along the banks of a small stream which flows 20 feet below their upper surface.[†] They consist of white marl, sand, and clay, in which are great numbers of, at least, two species of shells in a blanched and semi-fossil state, viz., *Melania tuberculata*, Müll., and *Melanopsis Saulcyi*, Bourg. These shells were found at several spots, as we wound our way through the channels which had been cut down through the horizontal strata of the marls.[‡]

Some distance north of our camp of the 12th December, I made the following section of the same beds :

SECTION OF THE LACUSTRINE STRATA NORTH OF ABU WERIDEH.

1. Fine gravel, evenly bedded	-	-	-	5 feet thick.
2. Fine gravel with beds of sand and marl	-	-	7	,,
3. Gravel	-	-	-	-	2	,,
4. Bed of coarse gravel with large stones	-	-	1½	,,
5. Coarse gravel	-	-	-	3	,,
		Total	-	-	18½

---

[*] See section of the Wâdy Arabah.—'Mount Seir,' p. 222.

[†] The terraces have no connection with the little stream, which has worn a channel through them, thus exposing the successive strata.

[‡] Two additional species of *Melanopsis* have since been identified by Mr. H. C. Hart, viz.: *M. buccinoides*, Oliv. ; and *M. eremita*, Trist.—'Quart. Stat.,' P. E. F., Oct., 1885, p. 264.

These beds of marl, sand, and gravel form a plain, with a gentle slope northwards towards the edge of the Ghor, spreading over a wide area, bounded on the east by the ridge of Samrat el Fiddân, and on the west by the limestone cliffs of the Tîh. Their southern limit I estimate at about 100 feet above the level of the Mediterranean, and they are traversed throughout a distance of several miles by the River Jeib, till they break off in an abrupt cliff along the southern margin of the Ghor, at a height of about 700 feet above the surface of the Salt Sea.

On Saturday, 15th December, we had again an opportunity of examining these deposits, in a position about 1,050 feet above the Salt Sea, and in a part of the great marly plain about 7 miles from the edge of the Ghor. Sections through the strata to a depth of 20 or 30 feet are here laid open. They consist of white calcareous marls, sometimes oolitic and pretty hard, alternating with beds of fine sand. At this spot, however, we could find no shells, but only fragments of plants incrusted with lime.

Wâdy Butachy.—This deep gorge, which breaks through the marginal cliffs of alluvial material which line the depression of the Ghor along the south, exposes a fine section of the old lacustrine deposits along its banks, and through a vertical depth of about 400 feet. Here the sloping plain suddenly breaks off; and near its eastern margin the deep channels of the Butachy and Gharandel issue forth from their mountain fastnesses.

The section of the Wâdy Butachy shows an intervening terrace between the upper surface of the plain and the bed of the existing stream, corresponding to a terrace which may be observed along the southern border of the Ghor itself. As far as I could judge, the banks are cut through two distinct sets of strata. The upper through a depth of 200 feet consisting of gravel and sand, the lower of white marls and clays. The change from the one variety to the other is quite distinct, and indicates some change in the physical conditions of the district during their deposition.

(*b*) *The Terraces of the Ghor.*—These terraces have been recognised by several travellers,[*] and are remarkably conspicuous. They occur on both sides of the Ghor and the Jordan Valley, as well as along the

---

* Lieutenant Lynch, 'United States Expedition,' 1849, p. 272. Tristram, 'Land of Israel,' p. 298, etc. Lartet, *supra cit.*, p. 6.

southern margin; and by their lighter colouring and nearly flat upper surfaces are easily distinguishable from the more ancient formations. One essential difference, however, is to be observed between the terraces on the east and those on the west of the depression ; for, while those on the former side consist of beds of sand and gravel, those on the opposite side consist, to a very large extent, of beds of white calcareous marl. The cause of this difference in composition is easily to be explained, and will be found in the nature of the formations on each side of the Ghor, from which the materials of the terraces have been derived. Those on the west side, being almost entirely formed of limestone, have given rise to calcareous deposits, either detrital or precipitated from solutions ; while those on the opposite side, being formed of various rocks (limestone, sandstone, trap, etc.), have given origin to beds of gravel and detritus containing less calcareous matter (Fig. 14). Still, the terraces are always lighter in colour and softer in texture than the bordering mountain ranges.

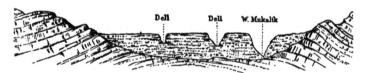

FIG. 14.—OLD LAKE BEDS OF GRAVEL ALONG THE EAST SIDE OF THE GHOR NEAR ES SAFIEH, FILLING IN OLDER VALLEYS, AND CUT THROUGH BY THE PRESENT STREAM.

Jebel (*Khashm*) *Usdum.*—Amongst all the terraces, of which Tristram has noticed no fewer than seven at Engedi, the most prominent and constant is that which has an average upper surface of 600 feet above the surface of the Salt Sea. This terrace includes the southern rim of the Ghor, and the isolated hill of Jebel Usdum, or the Salt Mountain, on the western shore of the Salt Sea, of which I have already given some account.* This remarkable terraced hill runs along the western base of the Judæan Hills for a distance of 7 miles, with an average breadth of 1¼ miles. The side next the sea breaks off precipitously, and exposes the structure of the whole mass ; while the interior is seen to be traversed by many channels and innumerable fissures, so that the mount appears as if

---

* 'Mount Seir,' p. 129 *et seq.*

about to fall to ruins. Seen from the opposite side of the Ghor near Es Safieh, Jebel Usdum appears to have a nearly level top, breaking off along a scarped face, and rent by a thousand fissures ; its lighter greyish or yellowish tints contrasting with the dark background of the limestone hills of Judæa. The upper plateau has a general level of 600 feet above the Salt Sea surface.*

The beds composing Jebel Usdum consist of alternating calcareous marls and shales, with selenite, forming the upper portion of the mass, and resting on laminated rock-salt, below which are shale and sandstone with beds of pebbles (Fig. 15). The following section of the lower strata was taken at about the centre of the hill opposite the ford across the River Jeib.

SECTION OF JEBEL (KHASHM) USDUM.

1. Beds of rock-salt, gypsum and marl   -   -   } *about* 110 feet thick.
2. Thin-banded rock-salt, with bands of marl and gypsum -
3. Irregular bed of pebbles, of slate, grit, and quartzite   -   ,,   3   ,,
4. Rock-salt   -   -   -   -   -   ,,   2   ,,
5. Bed of pebbles, often angular   -   -   -   ,,   2   ,,
6. Bed of brown shale, fine sandstone, and reddish sand, with
     pseudomorphs of salt crystals   -   -   -   ,,   30   ,,

In some places the rock-salt is more solid and massive than in others ; and at a spot about 2 miles from the northern end of the mount the salt is seen to be traversed by vertically bent joint-planes, along which large masses break off, and expose the base of the cliff, as shown in the drawing accompanying the narrative of the expedition, 1883-84.†

There can be little doubt that this terrace consists of a mass of lacustrine strata, formed at a time when the waters of the Salt Sea rose more than 600 feet above their present level. The terrace, though some-what isolated from the other terraces, corresponds, as regards its level, composition, and general aspect, with those which form the southern margin of the Ghor, except that rock-salt is here more largely developed as a component of the mass.‡ With all this evidence in favour of its

---

* I have already related Messrs. Hart and Laurence's venturous exploration of the upper surface of this remarkable terrace. They determined the level by aneroid observation. 'Mount Seir,' p. 131.

† Ibid., Fig. 17, p. 130.

‡ Major Kitchener, who examined the southern end of Jebel Usdum, told me that it was clearly connected with the lacustrine terrace towards the south of the Ghor.

lacustrine origin, it is hard to understand how so able an observer as M. Lartet could have come to regard the mount as formed of Nummulite limestone.* Its general relations to the older formation will be understood from the longitudinal section No. 2 and the following section (Fig. 15):

FIG. 15.—SECTION THROUGH JEBEL (KHASHM) USDUM.

1. Rock-salt resting on conglomerate.
2. Calcareous marls and shales with gypsum.
3. Recent terrace of the Dead Sea.

Before passing from this subject I may notice a remarkable structure in the rock-salt bed. When riding along the base of the cliff near the northern end of the hill, I observed the rock-salt to be traversed by oblique planes sloping southwards at an angle of about 20°, and overlain by white calcareous marls in a horizontal position. I was unable to determine, from this hasty view, whether the planes were those of bedding or jointage.

The Lisan.—No more remarkable object is to be seen anywhere around the shores of the Salt Sea than the Lisan; or 'tongue' of white strata which projects prominently forward from the eastern bank of the Ghor into the lake, and divides its waters into two very unequal portions.† It consists of a terrace, rising about 300 feet above the surface of the waters, of white chalky marl in thin beds, and with crystals and bands of gypsum‡ the surface sloping inwards, and connected with the coast by a low tract of slime and mud. According to Tristram, the neck of marl, where it leans against a spur of the Moabite Mountains, rises till it reaches

---

* Lartet drew this conclusion from observing the rounded blocks of limestone which strew the shore of the Dead Sea, and which he supposed had fallen from the top of the plateau.

† The best representation of this feature is given in the map accompanying Canon Tristram's 'Land of Israel' (p. 369), on a scale of 3 miles to 1 inch.

‡ Lartet, *loc. cit.*, p. 176.

an elevation of 500 or 600 feet, which is that of the principal terrace of the Ghor. The line of the terrace points in the direction of Khashm Usdum, with which it may be supposed to have been originally continuous, and to have had a similar origin.

(c) *Terraces of the Jordan Valley.*—The terraces which line the Jordan Valley above the head of the Salt Sea have attracted the notice of several writers, and have already been referred to by the author.* Dr. Tristram has identified seven such terraces at Engedi, and again near Jericho, of which the highest of alluvial deposits is stated to be 750 feet above the surface of the Salt Sea. Those I observed in this neighbourhood, extending from the banks of the Jordan to the base of the escarpment of Jebel Karantul, were as follows :†

Upper Terrace, with an elevation of 630 to 600 feet.
Second Terrace,   „   520 „ 250 „
Third Terrace,   „   200 „ 130 „
The alluvial flat, liable to floods,   90 „ 20 „

All these terraces, which are formed of beds of gravel, silt, and marl, slope towards the banks of the Jordan, so that the upper surface of each varies in elevation at different points, as might be expected in the case of successive lake-beds. At the same time they terminate along their lower margins in well-defined slopes and scarps ; and were a geological survey to be made of this valley in detail, it would be interesting to trace on a map the margin of the successive terraces.

The scarps and banks by which the terraces are terminated along the south end and sides of the Ghor indicate pauses in the process of falling away of the waters, and the action of waves and streams which have worn back the beds of alluvial material from the interior towards the margins. We may suppose that the beds themselves rapidly sloped away from the sides towards the deeper portions of the great gulf; and that during one of the pauses above referred to, when the waters of the lake had fallen so low as to lay dry the higher parts of the alluvial bed, waves and torrents acting upon the shallow and exposed portions would

---

* Lynch, 'United States Expedition,' p. 272. Lartet, 'La Mer Morte,' p. 176, etc. Tristram, 'Land of Israel.' Also, 'Mount Seir,' p. 161, etc.

† Determined by aneroid.

wear them back into cliffs. The materials thus dislodged would be carried into the deeper parts of the gulf and help to fill them up. In this manner, we may suppose, the section of the Dead Sea to the south of El Lisan, which is exceedingly shallow, was silted up.

It is well known that these terraces of alluvial materials extend up the Jordan Valley, on both sides, as far as the great plain which stretches to the north and west of the Lake of Huleh, and which marks the highest level of the great Jordan Valley Lake, corresponding in level to the extreme southernmost lake-beds of 'Ain el Weibeh in the Arabah Valley; both being a few feet above the level of the Mediterranean Sea. With this also corresponds the terrace of gravel described by Dr. Lortet near Safed, and already referred to.* No fact in physical history has been more clearly established than that the waters of the Jordan Valley Lake originally had a level somewhat higher than that of the Mediterranean, and considerably over 1,300 feet above its present surface.

(d) *Old Lake Beds in the Peninsula of Sinai.*—The existence of former lakes in the Sinaitic Peninsula has been recognised by Mr. Bauerman and the officers of the Ordnance Survey. Along the route taken by the Expedition of 1883, deposits of lacustrine nature were observed at intervals from Wâdy Wardan southwards to the Wâdies Hamr and Useit, and again in the Wâdy es Sheikh. The former were of large dimensions, being bounded on the east by Jebel Wutâh, on the south by Sarabût el Jemel, and west by Jebel Hammam Faroun. The beds consist of low terraces of soft laminated sandstone, clay, and gypseous marls, in a nearly horizontal position. In some places, beds of sand, gravel, and loose conglomerate replace the beds of marl.

The lacustrine strata of Wâdy es Sheikh somewhat resemble those of the Loess of the Rhine. They consist of terraces of stratified sand, fine gravel and marl rising from 60 to 80 feet above the present valley bed. They might be taken for old river gravels, were it not for the scarcity of large stones, and the fineness of the materials generally. The terraces in Wâdy Feiran, noticed by Bauerman and Oscar Fraas, are of similar materials, and contain beds of calcareous tufa, with sub-fossil shells, including *Lymnæa truncatula*, and a species of *Pisidium*.†

---

* *Supra*, p. 15.

† Mr. Bauerman contests the view of Dr. Fraas that these are glacial beds. 'Quart. Journ. Geol. Soc.,' vol. xxv., p. 35.

In the Arabah Valley, a considerable lake of shallow water appears to have lain in the hollow, bounded on the north by the ridge of the watershed, and to the south by a lower ridge, principally of gravel overlying limestone, which crosses the valley just south of latitude 30° N., at the base of Nagb el Salni.

Shingle Beds (Beach of Jordan Valley Lake).—I have above described the strata which were formed over the bed of the great Jordan Valley Lake; but not less interesting are the evidences of the marginal conditions which may be noticed at intervals, notwithstanding the changes which have undoubtedly taken place over the surface of the country during the thousands of years which have elapsed since the waters shrunk back from their original level. These marginal conditions are indicated by cliffs and terraces cut out of the solid rock; but in addition to these are heaps of boulders and shingle which I noticed in several places, particularly near the western slope of that remarkable ridge of granite and porphyry called ' Samrat el Fiddân ' (or Fedan), which runs for several miles in a north and south direction along the eastern side of the Arabah Valley, near its northern end. (See map of the Wâdy el Arabah.)

The position of this bank of gravel and boulders (G in fig. 16) is between the ridge of Samrat el Fiddân on the one hand and the alluvial plain (T) on the other. Its level is about that of the Mediterranean, and 250 to

FIG. 16.—SECTION ACROSS THE OLD GRAVEL BEACH WEST OF SAMRAT EL FIDDÂN.

S. S. Sand-dunes.
G. Gravel of old Sea-beach.
T. Terrace of Marls and of old Jordan-Arabah Sea.

260 feet above the alluvial plain to the west. At first when I came upon this ridge (in company with Mr. J. Armstrong, on the 13th December, 1883) I supposed it to be one of porphyry or granite, of which the blocks and pebbles were the surface fragments; but what was my surprise to find that they rest on yellow limestone and variegated sandstone (Cretaceous

strata). On examination, the ridge was found to consist of angular and rounded fragments and boulders of granite, porphyry, basalt, etc., piled up somewhat like a moraine, and forming a ridge parallel to the side of the valley. They had not fallen from the ridge of Samrat el Fiddân, for a broad valley separated the two ridges. To suppose it to be a moraine of glacial origin, in such a place, was out of the question ; and after consideration I came to the conclusion that it was a beach, heaped up by the waters of the ancient sea when it stood at its original level. The above section will probably explain the relations of the several features better than any description (Fig. 16).

Sandhills of the Maritime Coast.—All along the coast of Palestine from the borders of Egypt there extends a range of sandhills of remarkable height, and of a breadth of several miles, near Lake Sirbonis ; the range is continuous with the sand-dunes of Lower Egypt bordering the line of the canal. The levels taken by the officers of the Ordnance Survey of Palestine show that these sandhills reach a height of nearly 200 feet in some places.

Impelled by the prevalent westerly winds, these immense masses of loose sand have a tendency constantly to encroach on the interior, and are doing so where not restrained by some natural or artificial obstacle. South of Jaffa, in the waste called 'Arab es Suteriyeh, the sandhills cover an extent of ground 4 miles in breadth and 7 miles in length to the banks of the Nahr Rubîn, beyond which they again set in. I have already referred to the disastrous effects caused by these ever-drifting sands all along the seaboard of Philistia.[*] They have already destroyed much excellent land, and buried ancient cities, such as Gaza and Askalon. There can be no doubt but that the course of this destructive agent could be arrested by planting trees and bent-grass ; and this is a matter requiring the immediate attention of the Turkish Government.[†]

In endeavouring to account for the origin and formation of the sandhills, we must recollect both the composition of the underlying strata, and the physical changes which the district has undergone.

It is not improbable that much of the sand which has accumulated in

---

[*] 'Mount Seir,' p. 145 *et seq.*

[†] At Beyrût much success has attended the planting of fir-trees in arresting the inward drifting of the sands.

southern Philistia, about El Arish, has been drifted directly from Africa. Major Kitchener's account of the district between Tel Abu Hareireh and Ismalia is very instructive on this point, as showing the strength of the prevalent westerly winds and their power in carrying sand in the direction of their course ;* but we can scarcely refer the existence of the sandhills north of this place to a similar origin, as the wind sets in directly from the Mediterranean. We must therefore look to another source, and that (as it appears to the author) is ' the Calcareous Sandstone of Philistia.'

This formation, which I have already described as occupying the sea-board of Philistia and Western Palestine as far as Mount Carmel, readily decomposes under the atmosphere into its original sand. With a soft sandstone formation lining the shore and forming the sea-bed, we may well suppose it would be disintegrated and raised into sandhills by the power of the winds along the old coast-line during the period of elevation ; and as the process of depression proceeded, and the sea advanced on the land, the sand would continue to be drifted by wind-force in the same easterly direction. This process, probably dating far back into Pliocene times, has gone on down to the present day, with the result of converting into desert a large area of once fruitful soil.

The origin which is here ascribed to the sandhills of Western Palestine is very similar to that to which the most recent investigations have referred the sands of the Libyan Desert.† Sandstone strata of Cretaceous, and more recent geological, periods prevail over the southern and central portions of the Sahara, and the decomposition of these rocks under the atmosphere has given rise to accumulations of sand which the prevalent winds have drifted northwards and eastwards. The view that these sands are necessarily the materials of the ancient sea-bed, before the elevation of the Libyan region from beneath its waters, appears to be no longer tenable.

The sand-dunes of the Arabah Valley are of no great importance; they were encountered at intervals for short distances near our camp of the 4th, 6th, 11th, and 12th December. The material was originally derived,

---

* 'Mount Seir.' Major Kitchener's Report, Appendix.

† Zittel, *loc. cit.*, pp. 19-21. M. Rolland thinks the quaternary sandstone has been the source of the Desert sands, not the Nubian sandstone, as suggested by Zittel; perhaps both views are correct.

in all probability, from the decomposition of the sandstone strata; but may have existed as drifting sand since the time of the emergence, when the sea stretched up the valley for a distance of some 15 miles above the head of the Gulf of Akabah. The material of these sandhills is beautifully fine and clean, and in the absence of bent-grass is easily drifted by the winds. It retains in great perfection, on its surface, the foot-prints of the numerous wild animals which roam over the Arabah Valley during the night.

PART III.

CHAPTER I.

TERTIARY VOLCANIC ROCKS.

INDICATIONS of volcanic activity are limited to a very few localities over the region embraced by this treatise, and may be said to be almost exclusively confined to the district around the Sea of Tiberias and the adjoining district of Moab,* where they have been recognised by Burckhardt, Irby and Mangles, Tristram, and Lartet. The rocks of some of the localities in the vicinity of Mount Hor and of Es Safieh near the southern margin of the Salt Sea, described as of Tertiary volcanic origin, appear to me rather to belong to the group already described as of great geological antiquity. (See *ante*, p. 36.)

Composition of Lavas.—Almost all the volcanic rocks of Tertiary age in the region now under description consist of basalt and dolerite, with their varieties, of which the components are augite, plagioclase (Labradorite), olivine, and titano-ferrite. They are rich in iron, and poor in silica. The highly silicated class of volcanic rocks, which includes trachyte, domite, andesite, etc., is (as far as is known) unrepresented, except in one locality presently to be described, Khan el Ahmar, between Jerusalem and Jericho.

(a) *Table-land of Moab.*—Considerable masses of basalt and dolerite are found at intervals capping the surface of the table-land of Moab, and descending the beds of the deep gorges which lead down to the shore of the Salt Sea. Their truncated edges and disconnected positions lead to the inference that they are only portions of extensive sheets which have issued forth from several vents, at a time when the valleys and gorges were already in existence; at the same time it is evident that the valleys have been considerably widened and deepened, and that the basaltic rocks have been largely eroded subsequent to their eruption.

---

* A locality indicated on the geological map, near the southern extremity of the Sinaitic Peninsula, as also at Scherm, has been noticed by Burckhardt as containing dark volcanic rocks; but doubt has been thrown on the determination by Lartet, *loc. cit.*, p. 184.

The following localities are worthy of special notice. Commencing at the south, several large tracts of dark lava are found at the head of the Wâdy es Safieh, surrounding a thermal spring.

Proceeding northwards, on arriving at Rabbath Moab, north of Kerak, the limestone plateau is covered with fragments of basalt, of which materials the ancient ruins have been partly constructed.

Of the same material a portion of Jebel Shihân, and of the plateau between the Wâdies Mojib (Arnon), and Haidan is also formed. On either side of the Mojib, the cliffs of columnar basalt are to be seen capping the beds of white chalky marl, while a large mass has descended into the Wâdy Haidan, between the cliffs of limestone and marl on either hand.[*]

The volcanic phenomena of the district around Jebel Attarus, and along the deep gorge of the Wâdy Zerka Maïn, are of great interest. This dome-shaped mount, which rises conspicuously above the general level of the country, is itself formed of limestone; but on either side extensive sheets of basaltic lava have been spread over the Cretaceous strata, though they have failed to surmount the more elevated prominences. They have also descended the ancient bed of the stream down to the water's edge of the Salt Sea, and are laid open in the grand cliffs excavated since the period of eruption by the heated waters which issue forth at intervals from the highest sources at the margin of the table-land down to those of Callirhoë. (See *ante*, p. 23.) On the western flank of Jebel Attarus, and at a point which commands a great part of the gorge of the Zerka Maïn, may be observed considerable masses of scoria, peperino, and basaltic breccia appearing to surmount the head of a modern stream of lava, which may be seen stretching towards the northeast. This stream descends at first towards the bed of the Zerka Maïn, which it crosses, reaching the right bank, which it keeps for some time; it then returns to the left bank, and forms a line of prismatic columns along the side of the stream, and ultimately descends towards the margin of the Salt Sea.[†]

The basalt of the Wâdy Zerka Maïn, according to Lartet, contains grains of peridot (chrysolite) visible under the lens. Other outbursts of

---

[*] Instructive diagrammatic sections of these valleys are given by Lartet, *supra cit.*, Pl. v., Figs. 1 and 6. The basaltic lavas of Moab were first described by Seetzen in 1807.

[†] Lartet, *supra cit.*, p. 187. Tristram, 'Land of Moab,' p. 236.

basaltic lava occur, one at Mountar ez Zara, to the south of the Zerka Maïn, and the other at Wâdy Ghuweir, near the north-eastern end of the Salt Sea. The lava from this outburst descends into the waters of the sea, amongst which it disappears from view ; in structure it is scoriaceous and cellular.

It may here be observed that notwithstanding the assertions of some writers, there are no lava-flows along the western side of the Ghor or of the Salt Sea.*

(*b*) *The Haurân and Jaulân.*—It is, however, in the region of the Haurân, the Ledja, and the Jaulân (Jebel Heish), and Et Tulûl, all lying to the east and north of the Lake of Tiberias, that the ancient volcanic phenomena assume their grandest proportions, and where the great sheets of lava, the volcanic cones, and the rugged outlines of the surface somewhat resemble the volcanic region of Central France. The district now referred to lies rather outside the limits I have proposed to myself in this work, nor have I had the opportunity of giving it a personal examination. I shall, therefore, content myself with a short sketch of its general features, as gleaned from various authorities, and an account of the relations of its volcanic rocks to those of more ancient date.†

According to Lartet,‡ the region referred to, called in the Bible ' The Land of Bashan ' or ' Trachonitis,' contains three great tracts of volcanic rocks, of which that called Et Tulûl lies to the east of Damascus, that of the Haurân and the Ledja to the south of this city, and that of the Jaulân extends from the Valley of the Hieromax along the eastern shore of the Sea of Tiberias and the Jordan Valley to the southern base of Mount Hermon, against the uprising spurs of which the sheets of lava have been abruptly terminated.

---

* Such as Rüssegger, who states that the western side of the Jordan Valley is penetrated by innumerable dykes and streams of basalt. It is probable he has mistaken the bands of dark chert or flint for this kind of rock.

† Van der Velde's description of supposed volcanic phenomena on the shores of the Bahr Lût, ' braune Lavabrocken, in lothrechten Wänden aufeinander gethürmt . . . dazwischen kraterförmige Hügel von weisser, gelber und grauer Farbe, Alles Erzeugnisse des unterirdischen Feuers,' is justly characterized by Fraas as ' reine Gebilde einer aufgeregten Phantasie und der geologischen unkenniss.'—' Aus dem Orient,' p. 65.

‡ Much of the following account of the volcanic rocks is taken from M. Lartet's ' La Mer Morte,' pp. 185-189.

The rocks of this region consist of augitic lava (basalt and dolerite) often vesicular and scoriaceous, and they spread over large tracts of country from the orifices or craters of the interior, which rise conspicuously above the general sheets of lava. One of these streams, described by Russegger, issues from a mountain in the Jaulân, at a height of 955 feet above the Lake of Tiberias, and flows to the very borders of the lake, in a current no less than a league in breadth at its termination. The lava is partly compact and partly cellular, containing much zeolite.

The stream of the Hieromax, which flows into the Jordan just where that river issues from the Lake of Tiberias, has cut its course through the lava, and flows sometimes between walls of basalt on the one hand, and of limestone on the other. (See note, p. 99.)

The whole of the eastern shore of the Sea of Tiberias is covered with basaltic *débris*, and many lava-floes descend below the waters of the lake itself, as, for example, at the outlets of the Wâdies Sik and Es Semak. Near this latter valley the basalt is of a greyish colour, containing red grains of altered peridot. This rock is vesicular, and in the cells are crystals of carbonate of lime.

Between the Sea of Tiberias and the base of Hermon at Bâniâs, the traveller marches for two entire days over sheets of basaltic lava. The region is sterile and forbidding in aspect. It is described by Ali Bey, who traversed it in 1807, as ' Etant une région d'un aspect infernal.'

The basalt of the Jaulân is generally compact, of a dark greyish tint, and containing numerous crystalline grains of peridot.

The volcanic rocks rise into the elevated tract of Jebel Heish, and on the summit of the plateau is situated Lake Phiala, an ancient crater. The line of conical tells which extend along this plateau towards the south, are, in like manner, all volcanic. The loftiest is Tell Abu Nida, south of Phiala, rising about 4,100 feet above the sea. It has a deep crater, thickly studded with oak trees. The adjacent Tell 'Eram is nearly as high, and has also a crater.[*] The decomposition of the basalt has given origin to the rich loam of the Haurân, celebrated throughout Syria.

At Bâniâs, a lava-stream descends into the valley, flowing round the

---

[*] Robinson, quoting Dörgens, in ' Berl. Zeitsch. für Erdk.,' Nov., 1860, p. 405-6.

base of Mount Hermon (Jebel es Sheikh), and spreading out on the plain of Ard el Huleh. The head streams from the Anti-Lebanon have cut for themselves a channel under the lava between Bâniâs and Tell el Kadi, and issue forth under the name of 'The Sources of the Jordan,' in copious fountains.*

It would appear that the lavas of Safed and of the Jaulân were originally connected, and had pent up the waters of the Jordan between the Lakes of Tiberias and Huleh, so that the waters of this latter once stretched over a much larger tract towards the north than at present. The tract called 'Huleh Marsh,' extending to El Kalisha and Zuk et Tahta, was then under water. Afterwards the Jordan cut a channel for its bed and the lake was partially drained and fell to its present level. Its surface is about seven feet above that of the Mediterranean.

(c) To the west of the Lake of Tiberias the most important mass of volcanic materials is Jebel Safed, which Rüssegger has regarded as a centre of eruptions of this region. The sheets of basaltic lava stretch southwards in the direction of the lake by Tell Hûm. And in Jebel Jish, which lies about 5 miles north-west of Safed, Dr. Robinson has identified an actual crater.† It is described as an oval basin, about 400 feet long, by 120 feet broad, and 40 feet deep; the sides are shelving, but steep and rugged. The basin is usually partly filled with water, and is known as Birket el Jish. Safed, it will be recollected, was visited by a terrible earthquake on New Year's Day, 1837, which destroyed the entire town, and in which four thousand lives were lost.

To the west of the Lake of Tiberias there occur several other volcanic masses. Amongst these a stream of basaltic lava, which appears to have had its origin in the hill called Kurn Hattin, rising 1,178 feet above the sea, has flowed eastwards to the borders of the lake near the village of Tiberias (Tubarieh) itself. The basalt contains grains of peridot, and crystals of zeolite, amongst which Rüssegger has indicated those of mesotype. Other basaltic localities are those of Cabul (Kabûl) Umm el

---

* M. Lartet has given a diagrammatic plan of these remarkable phenomena, which find their counterpart in the Auvergne and other volcanic regions. · Tristram has described the basaltic ridges of this district, 'Land of Israel,' second edition, pp. 425, 426, and 435.

† 'Physical Geography of the Holy Land,' p. 289. Anderson, 'Geological Report,' pp. 128—9. This latter author describes several other small craters in this neighbourhood.

Kaláid, Wády Maleh, Zer'ain* (or Zerín) and Abu Shushah,† near Ramleh.

(*d*) *Volcanic Rock of Khan el Ahmar.*—Canon Tristram has noticed an outburst of volcanic rock‡ at this place, and it is entitled to special description from the fact that it appears to be the only representative of the felspathic class of volcanic products yet observed.

On approaching the hill on which stands the Khan el Ahmar from the west, we observe the beds of limestone and chert to be much disturbed and contorted along the sides of the valley. The cause of this presently becomes apparent, when on crossing a brook-course, we find the limestone replaced by a red, yellow, and white felspathic rock, which in its general aspect resembles domite. As far as I could observe, this intrusive mass only just reaches the surface, and has so calcined the limestone that it is difficult to determine its true nature. Between the brook-course and the Khan, the felspathic rock seems to be coated with a thin layer of calcined and altered limestone, which conceals the former from view; but judging by the change which the latter has undergone, it may be supposed to lie close underneath this thin coating.

Geological Age of the Volcanic Eruptions.—The Tertiary age has been fruitful of volcanic products over large portions of the world, and within the regions bordering the Mediterranean. Those of Central France, of Italy and the adjoining islands, of the Grecian Archipelago, and of Asia Minor, may be specially named. The volcanic lavas of Syria and the adjoining districts are referable to a portion of the same prolonged period; but from their relations to the rocks and features of the Jordanic depression, it is clear that they are more recent than the Eocene epoch, and, at least, to a certain extent, than the Miocene.

Recollecting the manner in which, both in Moab and the Jaulân, the basalt streams flow along depressions hollowed out of Cretaceous and Eocene limestones, it is clear that the basaltic eruptions are of later date than the depressions themselves; and we shall probably not be in error if we assume that the earlier manifestations of volcanic action began

---

* 'Quarterly Statement Palestine Exploration Fund,' July, 1874, p. 186.
† Tyrwhitt Drake, 'Quarterly Statement Palestine Exploration Fund,' April, 1872, p. 42.
‡ 'Land of Israel,' p. 195.

during the epoch of the Pliocene. I have already hazarded the conjecture that the outbreak of volcanic action is indirectly connected with the filling up of the great Jordan Valley Lake during the Pluvial period. We know that the proximity of large sheets of water is indispensable to volcanic action. What more probable than that the waters of this ancient lake, of very considerable depth and pressure, penetrating the interior along the great line of fissure of the Jordan Valley and its branches, should have reached the internal heated masses, and set in action the subterranean laboratories, which afterwards gave such striking evidence of their proximity in the great sheets and streams of lava which have overflowed the regions on both sides of the Jordan Valley, and invaded the waters of the Lake of Tiberias?

POSTSCRIPT.—While these pages were passing through the press I received (through the kindness of Professor Roth, of Berlin) an interesting 'Preliminary Report' (Vorläufiger Bericht) on the geological structure of the country east of the Jordan, written from Haifa, by Herr F. Noetling. I look forward with interest to the full Report on this interesting district. With reference to the Valley of the Yarmuk (Hieromax), Herr Noetling states that two lava-streams are clearly visible—the older, which filled up the valley of the river originally, cut out of Cretaceous limestone ; this was, subsequently, partially re-excavated by the stream ; the younger, which flowed down over the bed of the stream thus formed, and which may be observed to rest upon the alluvial gravels of this second bed, containing specimens ot *Melanopsis*. Again, through this last lava-stream, the river has again cut its channel so deeply that it has entered the underlying limestone formation. All this would indicate a considerable lapse of time between the older and younger basaltic flows.

PART IV.

CHAPTER I.

DYNAMICAL GEOLOGY.

HAVING passed in review the various rocks and formations entering into the structure of the region now under consideration, we proceed to another part of our inquiry; viz., the course of events and the mode of operation of the agencies which have resulted in the production of the present physical features of this region. The subject comes under the head of dynamical geology; and in handling it we have to deal with terrestrial forces acting within the earth's crust on the one hand, and the various agents of denudation or erosion which have operated from without, on the other.

It will be inferred from what has been said, that down to the close of the Eocene period, the whole tract now being reviewed, together with the adjoining regions of Northern Africa, the Arabian Peninsula, and Syria, formed the floor of the ocean; the only possible exception being the higher elevations of the Sinaitic Mountains, and those formed of similar very ancient rocks on both sides of the Red Sea. Over the floor of this ocean, successive beds, chiefly of limestone, had been laid down during the Cretaceous and Eocene periods, until they had reached a thickness of several thousand feet; but when this latter period came to a close, and that of the Miocene commenced, deposition of strata was generally suspended, owing to movements which elevated out of the waters the present land areas, and depressed to still greater depths other portions of the same general sea-bed. It may be stated in general terms, that during the Miocene epoch, the general outlines and areas of land and sea were roughly marked out and determined; and, with more or less modification, have remained as such down to the present day.

Along with such outlines, other features, extending over the districts inland from the present seas, were also, in the main, defined. The mountain ranges of the Lebanon and Anti-Lebanon were upraised, and in connection therewith the remarkable linear depression, which I have called 'the Jordan-Arabah Valley,' which extends from the western flank of Mount Hermon in a nearly straight line southwards into the Gulf of Akabah, was formed. The general mode of formation of this leading physical feature it now becomes necessary to investigate.

(a) *Formation of the Jordan-Arabah Valley; Line of the Main Fault.* —That this deep depression is the direct result of a 'fault,' or fissure, of the crust, accompanied by a displacement of the strata, the relations of the formations on opposite sides leave no room for doubt. The fact has been recognised by several authors, from Leopold von Buch downwards;[*] and has been recently demonstrated in some detail by M. Lartet, with whose views on this point my own are in accord.

An examination of the geological maps accompanying this volume will show that there is a general dissimilarity in the geological structure of the opposite sides of the Jordan-Arabah Valley throughout the greater part of its extent from the head of the Gulf of Akabah to the plain of Jericho, north of the Salt Sea. It will be observed that the formations occupying the eastern side are older than those of the western; and that, generally, while the upper Cretaceous[†] limestones form the western borders, strata of Archæan, Carboniferous, or Cenomanian age occupy the lower sides of the eastern border ranges. But even when this dissimilarity of the strata does not appear, as along the line of the Jordan Valley between the plains of Jericho and the Sea of Gennesareth, the strata of the same formation on either side may be supposed to be different; those on the west being superior to those on the east.

This general dissimilarity in the stratification indicates a displacement along a line of fault, to an extent of several thousand feet in some cases; the maximum displacement being at the southern end of the Ghor, opposite Mount Hor, and for some distance south. The effect will be easily understood on referring to the longitudinal sections (Nos. 2 and 4), which have been drawn at several points across the Great Valley.

---

[*] For example, by the late Dr. Hitchcock, 'Rep. Assoc. American Geologists,' Boston, 1843, p. 369. Professor W. W. Smyth, Anniv. Address Geol. Soc. Lond., 1868; 'Quart. Journ.,' vol. xxiv. Dr. Tristram, 'Land of Israel,' second edition, p. 320, etc.

[†] It is probable, also, that Nummulite limestone of Eocene age forms in some cases the western borders of the Great Valley.

Owing to the covering of sand, gravel, or of ancient lacustrine strata over the floor of the Arabah, the actual fault or fracture is not often visible; but its presence is generally indicated by the sudden uprise of the older strata on the eastern side. Between the Gulf of Akabah and the Ghor, the main fault follows the eastern side of the Valley, at the base of the Edomite Mountains, and its actual position can be determined (1) at the base of Jebel esh Shafeh opposite the Wâdy Redadi, where the limestone is broken off against the granite and porphyry; (2) at the entrance to the Wâdy Gharandel, as described in 'Mount Seir';* (3) just at the watershed of the Arabah Valley, where the limestone is again seen to be broken off against cliffs of porphyry; (4) at several points farther north along the same line, as at Nagb er Ruhai, and again at the northern end of Samrat el Fiddan. (Fig. 17) North of this point the line of fault is concealed by recent deposits.

FIG. 17.—SECTION THROUGH NORTH END OF SAMRAT EL FIDDAN.
Beds of Cretaceous Limestone and Marl resting on Sandstone. Porphyry traversed by dykes.

(*b*) *Secondary Faults of the Jordan-Arabah Valley.*—While there is one leading and continuous line of dislocation along the Jordan-Arabah depression, numerous minor fractures also occur throughout. Some of these are sufficiently large and important to admit of being inserted on the geological maps; and their effects on the stratification will be better understood by a study of the maps themselves, than by any attempt at description in detail.

Some of these faults run in a line parallel to that of the main fracture— such as that which I have described in 'Mount Seir' (p. 66, with Fig. 7)

---

* Page 84; the fault is accompanied by a spring of water, as described by Dr. E. G. Hull.

—which occupies the Wâdy el Musry, and enters the Gulf of Akabah on its western shore, opposite the Island of Jeziret Faroun; or the large fault which runs north and south at the western base of Mount Hor (Fig. 19).

FIG. 18.—SECTION THROUGH MOUNT HOR.   (J. HAROUN.)

At the right of the figure Mount Hor, formed of beds of red sandstone and conglomerate, is seen resting on a mass of granite and porphyry traversed by dykes; against this beds of Cretaceous Limestone are thrown down by a large fault on the left.

Others branch in various directions from the main fault; such as that which, branching off in a south-south-east direction, strikes into the Edomite Mountains, producing successive terraces of the sandstone beds surmounting their granitic bases; and the large fault which enters the Ghor from the south along the Wâdy Gharandel, and which displaces the strata in the Jordan Valley north of Kerak. Together with these

FIG. 19.—FAULT IN WÂDY ET TIHYEH.

P. Porphyry, etc. (old fundamental rocks).
S. and L. Cretaceous Limestone and Sandstone.
L. Limestone and Marl with beds of Chert.

may be mentioned the faults which break the continuity of the escarpment of the Tîh plateau on approaching the Gulf of Akabah. One of these is worthy of special notice from the important effect it has on the physical geology of the district now referred to. Its direction is due north and south; and, therefore, nearly parallel with the great Arabah Valley fault. It is accompanied by at least another in a parallel direction, about 4 miles farther to the eastward. It occupies the line of the Wâdy el 'Ain for some distance, and going northwards, crosses into the Wâdy et Tihyeh, and ranges along the western base of Jebel Aradeh, having a downthrow to the eastern side. During our travels through this region, we were able to notice the effect of this fault for three successive days. The above sketch (Fig. 19) is taken in the Wâdy et Tihyeh.

(*c*) *The Jordan-Arabah Fault an Axis of Disturbance.*—Whatever may have been the cause of the terrestrial disturbances along the special line of the Jordan-Arabah depression, it is sufficiently clear that this line was an axis of disturbance for the whole region now under consideration. Along this line the strata are either displaced by secondary faults, or contorted and tilted at various angles; while, as we recede from the line of the displacement, the beds generally begin to assume a position approaching the horizontal. Travellers who have made excursions into the table-land of Edom and Moab all concur that the strata are nearly horizontal; and my own observations, though not very extensive in this direction, tend to confirm this view. From what we can learn, the beds over the great Arabian Peninsula are also but slightly inclined.

Extending our observations in the opposite direction, the same statement largely applies. As far as we know, the strata over the table-land of the Tîh are (with occasional exceptions) but little removed from the horizontal position.[*] The Gulf of Suez seems to occupy in its upper part the axis of an anticlinal arch, and in its lower the line of a fault; and fault-lines traverse the Wâdy Nasb and Jebel Attaka; but with these exceptions, the strata appear to be generally undisturbed.

Crossing into Northern Africa, there can be no doubt that the line of the Nile Valley from Cairo southwards coincides with a line of displacement, as Sir J. W. Dawson has pointed out,[†] whereby the strata of Jebel Mokattam are brought down to a lower level on the opposite side of the

---

[*] The exceptions being mainly in the district of the Wâdy Nasb and Wâdy Hamr.

[†] 'Geol. Mag.,' July, 1884.

valley; but with this exception, all the beds of the Cretaceous and Tertiary periods over the Libyan and Nubian Deserts are remarkably undisturbed and unbroken. Thus it appears that it is only when we approach the position of the great rent in the earth's crust which follows the line of the Jordan-Arabah depression that evidence of powerful dynamical action becomes strongly manifest; and that we are brought face to face with the result of those forces, acting within the interior of the earth's crust itself, which have originated the leading features of the surface.

(*d*) *Mode of Formation of the Jordan-Arabah Depression and of the Basin of the Dead Sea.*—Having thus pointed out the evidence for the existence of the great rent and displacement of the strata along the line of the Jordan-Arabah depression, we have to consider the process, or *modus operandi*, by which the depression itself was produced, together with the formation of the elevated table-lands on either hand.

In the first place, we must dismiss from our minds any idea that these movements in the earth's crust were spasmodic or cataclysmal. Speaking generally, it is far more probable that they were gradual, and that they extended over a long period of time.

I am disposed to think that the fracture of the Jordan-Arabah Valley and the elevation of the table-land of Edom and Moab on the east, and of Palestine on the west, were all the outcome of simultaneous operations, and due to similar causes, namely, the tangential pressure of the earth's crust due to contraction.[*]

If this be so, we may suppose that at the close of the Eocene period the long era of repose, and of deposition of strata, gave place to one of movement and rupture of the strata; and, owing to powerful lateral pressure acting from east and west directions, the strata over the whole area now under consideration were forced into a series of synclinal and anticlinal curves, at right-angles to the direction of pressure, which are represented and replaced by actual fractures and displacements where there were lines of weakness; such lines of weakness are those of the Jordan Valley and the Valley of the Nile.[†]

---

[*] The contraction being in its turn due to the secular cooling of the crust.

[†] If we begin at the east (*a*) the Jordan Valley may be considered as a broken synclinal or trough; (*b*) the table-land of Central Palestine as an anticlinal (or arch); (*c*) adjoining Mediterranean bed, extending southwards into the central plateau of the Tíh, as a low

Upon such principles we may conclude that, as the land area was gradually rising out of the sea, the table-lands of Judæa and of Arabia were more and more elevated, while the crust fell in along the western side of the Jordan-Arabah fault; and this seems to have been accompanied by much crumpling and fissuring of the strata. The great arches, now forming table-lands, would be the first lands to appear above the surface, while the intervening hollows (inverted arches or synclinals) would be occupied by sea-water. Owing to this cause it is probable that the first waters of the Salt Sea were those remaining from the ocean itself. The fauna of the Lake of Tiberias may thus have had, in the main, a marine origin.

Confining our attention now to the trough of the Salt Sea (or the region of the Ghor), we may suppose that as the lands on either hand rose, the bed of the trough, owing to continued subsidence, became more deep over the area now occupied by the waters of the Salt Sea itself; and into this gulf all the waters flowing from the bordering lands would naturally empty themselves.

Origin of the Faunas of the Sea of Tiberias and of the Jordan.— In this connection I am tempted to offer a few observations on the mode by which the fishes and molluscs of the Sea of Tiberias may have originated. For a knowledge of these remarkable forms we are indebted to several authors, of whom Roth, Locard, Lortet, and Tristram may be specially named.[*] The abundance of fishes in these waters is known from both sacred and secular history, and is still testified to by statements of travellers; but, considering that these waters are land-locked by broad and high barriers of land from communication with either those of the outer sea or of neighbouring lakes and streams, it may well excite surprise how they have been stocked with living forms except by the hand of man, a supposition quite inadmissible.[†]

As regards the molluscs, the genera to which they belong are of wide distribution; but the species are for the most part peculiar, no less than

---

synclinal; (*d*) the Isthmus and Gulf of Suez as an anticlinal; the range of hills between the gulf and Nile Valley as a low synclinal; (*e*) the Nile Valley a broken anticlinal.

[*] Dr. Roth, 'Spicilegium Molluscorum,' 1855; and 'Coquilles Terrestres et Fluviatiles,' ed. by Mousson, 1861; Arnould Locard, 'Malacologie du Lac de Tibériade,' 1883; Professor L. Lortet, *supra cit.*, p. 6; Dr. H. B. Tristram, *supra cit.*, p. 6.

[†] Chiefly on the ground that some of the forms are distinct and peculiar.

sixteen of *Unio* being confined to the Jordan and its feeders ;* while
from the sub-fossil forms found in the terraces of the Jordan Valley, it is
clear that most of them have descended from Quaternary times. As
regards the fishes of the Jordanic basin, only one, viz. *Blennius lupulus*,
out of thirty-six species belongs to the ordinary Mediterranean fauna ;
two others, *Chromis niloticus* and *Clarius macracanthus* are Nilotic ; and
seven other species occur in other rivers of South-western Asia, the
Tigris, Euphrates, etc. Ten more are found in other parts of Syria, and
the remaining sixteen species are peculiar. Dr. Tristram considers this
analysis points at once to the close affinity of the fauna of the Jordan
with that of the rivers of tropical Africa ;† but in any case it points to an
original connection between the waters of the Jordanic basin and those
of the rivers entering the adjoining seas.

The origin of fresh-water faunas has recently engaged the attention of
naturalists : and to this subject Professor Sollas has added a valuable
contribution ;‡ in which he arrives at the conclusion that, as the con-
version of comparatively shallow continental seas into fresh-water lakes
has taken place on a large scale several times in the history of the earth,
this has been accompanied by the transformation of some of the marine
into fresh-water species. The most remarkable example in the Eastern
hemisphere is, perhaps, that of the Caspian, in which there is a strange
commingling of lacustrine peculiar forms with others of oceanic origin,
due to a connection with the Arctic Ocean which has since been cut off.
The peculiar forms may be regarded as evolved out of those which existed
when the Aralo-Caspian area was more or less connected with the outer
ocean. To a similar origin may we not assign the great majority of the
Jordanic forms ? We have seen that during the progress of formation
of the great Jordan-Arabah depression, while the lands on either side were
emerging from beneath the waters of the Miocene Sea, there is reason
for believing that the waters of the outer sea were caught up and retained.
(See p. 109.) If so, it follows that the animate forms inhabiting the
waters were also enclosed, and ultimately became isolated (so to speak)

---

* Tristram, *loc. cit.*, p. 178.

† Ibid., Preface, p. xii.

‡ 'On the Origin of Fresh-water Faunas ; a Study of Evolution,' Scient. Trans. Roy.
Dublin Soc., vol. iii. (ser. ii.), p. 87 *et seq.* (1884).

within the limits of an inland basin which they were never to leave. Many marine forms, such as corals, crinoids, etc., would perish, but a process of transformation would commence and proceed in the case of many of the species of fishes which, going forward through the long lapse of Pliocene and Pluvial times, would result in peopling the waters with those special forms which we now find living. For such a process those of fishes are peculiarly adapted.

If these views be correct, then the fishes of the Jordanic basin are the descendants of those which lived in the waters of the Eocene Seas. Their ancestors inhabited salt waters ; the descendants fresh. But observation and analogy have shown that not only are there some fishes which, like the salmon and sturgeon, are capable of a twofold habitat, but that (provided the process of change be sufficiently gradual) animals which belong properly to one variety of element may become naturalized to the other. The change from the salt water to the fresh in the case of the Sea of Tiberias and the Lake of Huleh has doubtless been one of extreme slowness during the time that the great Jordanic Lake was retiring into its present limits ; and thus the law of ' descent with modification,' which in the case of fishes receives a fresh illustration, has had ample time to come into operation.

(*e*) *Formation of the River Valleys.*—This brings us to the consideration of the formation of the physical features, but especially of the river valleys of the region now under review. We may assume that from the moment the lands appeared, the streams originating in the rainfall began to flow along their easiest courses towards the outer sea on the one side and the Jordan-Arabah depression on the other; and each stream, with its tributaries, being once established, would continue to deepen its channel as time went on. In this way the present drainage system of Palestine, Moab, and Arabia Petræa was originally established, and appears to have assumed very definite proportions even as early as the Miocene or early Pliocene period ; for, from the relations of the ancient terraces of the Jordan Valley to the depression of the Ghor, it is clear that the principal ravines and deep watercourses which debouche on the Jordan Valley had been excavated before the terraces themselves were formed.*

---

* These more ancient terraces may be considered as dating as far back as the Later Pliocene period, and as being contemporaneous with the second partial submergence of the land.

At this epoch, namely, the close of the Miocene or commencement of the Pliocene, the waters of the Dead Sea had contracted to at least their present level, for we find these old river-channels (now partially filled in with lacustrine deposits) worn down in the older formations to the margin of the existing lake, and this could only have been effected if the slopes were dry land ; for, under the waters of a lake, river-erosion cannot be carried on.

Thus, at the close of the Miocene epoch, we behold the existing land surfaces on either side of the Jordan-Arabah Valley in a condition not very different from that of the present day, at least in their main features. Throughout that prolonged period, terrestrial forces on the one hand, and atmospheric agencies on the other, were engaged in the formation of ridge and furrow, of hill and valley, until in its rough outlines the whole region had assumed somewhat of the aspect it bears at the present day ; namely, that of successive table-lands deeply intersected by numerous ravines and winding valleys.

The epoch during which the waters of the Salt Sea sank down to, or below, their present level may, with much probability, be correlated with that during which the Mediterranean waters were so reduced in volume that this great sea seems to have been converted into a chain of lakes, owing to which many of its islands, such as those of Malta and Sicily, became united to the mainland, and were the abode of elephants, hippopotami, and fresh-water turtles. As the question is put by Mr. Jamieson : ' In what other way can we account for foxes being found in Minorca ; hares, martens, deer, foxes, etc., in Corsica and Sardinia ?'* A general lowering of the Mediterranean waters is an inference deducible from the distribution of the animals, either recent or extinct, throughout its islands, at an epoch not very remote, and which may be represented by the Interglacial stage of the Quaternary period ; the process of dissection may have been extended to the waters of the Salt Sea.

---

* 'Geological Magazine,' May, 1885, p. 199. On this subject the reader is referred to the papers of Captain Spratt, 'Quart. Journ. Geol. Soc.,' vol. xxiii., p. 283 ; Professor Ramsay and J. Geikie, ' On the Geology of Gibraltar,' ibid., vol. xxxiv., p. 505.

CHAPTER II.

CONTINUING our review of the changes which the region underwent during the period intervening between the Miocene and the present, we arrive at an epoch which has left its marks specially over the maritime and Jordan-Arabah districts. The Pluvial epoch was one which was characterized by a general subsidence of the whole region bordering the Levant, and by the rising of the inland waters till they had converted the great Jordan-Arabah depression into a lake over 200 miles in length, and over 2,000 feet in depth.

The evidences of the subsidence along the coast of Palestine, Syria, and Northern Africa have already been offered (see *ante*, p. 68 *et seq.*), as also have those of the former extension of this great lake (p. 87 *et seq.*). It is only now necessary to discuss the causes which led to the remarkable increase of water in the tributaries of the Jordan Lake.

It is probable that the mere subsidence of the land to an extent of 200 to 250 feet, resulting, as would be the case, in the submersion of a considerable tract which had previously existed as a highly heated land-surface, would of itself have tended to add to the humidity of the atmosphere, and to increase the rainfall. The effect of laying under water the whole of Lower Egypt, the Nile Valley, the Isthmus of Suez, and adjoining low tracts, would necessarily produce some effect of this kind, but scarcely to the extent required to account for the swelling of the waters of the Jordan Valley Lake to their highest level of 1,300 or 1,400 feet above their present surface.

In searching for a further explanation we cannot fail to recollect that the Pluvial epoch includes that of the Post-Pliocene or Glacial; one marked by an extraordinary refrigeration of the temperature of the northern hemisphere, and one which, in the present sub-tropical regions, must have been marked by an amount of rainfall similar to that which is now precipitated over the temperate zone.

At the present day, and in the Old World, perennial snow and glaciers are restricted to the Mountains of Scandinavia, to the Alps, the Pyrenees, the Caucasus* and the Himalayas; but during the Glacial epoch they were spread over nearly the whole of the British Isles, the northern parts of Europe and Asia, as far south as latitude 52° N.; while amongst those mountain-chains above named, the snows and glaciers were of greater depth and vastly greater extent than at the present day.†

Amongst the districts where perennial snow and glaciers existed at this epoch were those of the Lebanon and Hermon in close proximity to Palestine and Arabia Petræa, and the effect of this proximity on the climate of the latter can scarcely be doubted.‡ The presence of moraines and other glacial phenomena in the Lebanon has been pointed out by Sir J. B. Hooker,§ and more recently by Canon Tristram and M. Lartet. Amongst the more remarkable moraine accumulations are those of the Kedisha Valley, described by Hooker. Near the head of this valley is the forest of venerable cedars, growing on an ancient moraine which crosses the valley at an elevation of about 4,000 feet above the sea, and divides it into an upper and lower floor. The cedars grow on a portion of the moraine which borders the stream, and are doubtless but a remnant of extensive forests of the Glacial epoch; they are also to be met with in other parts of the Lebanon range. It is not improbable that at the time when the glaciers descended into the

---

* The glaciers of the Caucasus, which are now several miles long, were once more extensive, probably during the Glacial epoch. According to M. Dinnik, the modern glaciers have decreased in size considerably during the present century.—'Mem. Caucasian Geographical Society,' vol. xiii.

† See Mr. Andrew Murray's map, No. 4, in the 'Geographical Distribution of Animals,' and the glacial map of the British Isles in my 'Physical History of the British Isles,' Plate. xiii. (1882).

‡ Oscar Fraas supposes there were glaciers amongst the Mountains of Sinai, and points to the presence of moraines and erratic blocks in the Wâdy Feiran, etc. ('Aus dem Orient,' pp. 28-30, etc.), but more recent observers do not concur in this view; and for myself, I saw no decided evidence of former glacial conditions when in this district.

§ 'On the Cedars of Lebanon,' etc., 'Nat. Hist. Review,' 1862, p. 11.

valleys of the Lebanon, the mean annual temperature was 25° Fahr. lower than at present over Palestine and Syria.*

With a climate in Palestine and Syria during the Pluvial period resembling that of the British Isles, we may well suppose that the Jordan was a larger stream than at present, and that its numerous affluents were copiously supplied from the rainfall. Hence it is not surprising that the waters in the Jordan Valley Lake increased in volume, and gradually rose till at last they reached the level of the outer sea, and stretched nearly from the base of Hermon southwards into the Arabah Valley. There is no evidence, however, that they ever overtopped the watershed, or, indeed, reached within several hundred feet of its upper surface. From the time the land was raised out of the sea, and the Jordan-Arabah depression became a physical feature of the region, the low ridge which crosses the valley near the base of Mount Hor appears to have been dry land.

We may infer that towards the close of the Pluvial epoch the waters of the inland lake reached their highest level; and that as the glaciers and snows disappeared from the Lebanon, and the more modern physical conditions set in, the rainfall became less in amount, and the surface of the great lake gradually fell away, until the Jordan Valley became the bed of two lakes of comparatively small dimensions, and of a connecting stream.†

To the Pluvial epoch must, also, be referred the erosion of the deep gorges and ravines of Palestine, and of the still wider valleys of Arabia Petræa, which are seldom otherwise than dry beds of former rivers. The general desiccation above referred to appears to have affected the climate of extensive regions bordering the equator, both in Asia and Africa.

That Northern and Central Africa has undergone a remarkable change in the climate, even within comparatively recent times, can scarcely be doubted. Districts which at the commencement of the Christian era

---

* The mean annual temperature of Beyrût is about 70° Fahr. During the Glacial epoch it may have been as low as 40° or 45° Fahr.

† To this epoch may be referred the extension of the waters of the great lakes of Asia, when those of the Caspian and Sea of Aral were united, and probably communicated along the depression of the river Obi with the Arctic Ocean. In this way we may account for the presence of seals, fishes, crustaceans and molluscs in the Caspian, which are identical with, or closely resemble, those of the Arctic Ocean.

were productive of grain, are now sterile deserts. The elephant, which requires abundance of water and of green food, has retreated southwards of the Soudan, and is replaced by the more enduring camel, an animal whose native home is in Asia. The elephant, in the time of the Romans and Carthaginians, existed in a wild state over Northern Africa; and its destruction there, though chiefly due to man, is not altogether so, but also to the absence of vegetable food and of cover. The illustrious traveller, David Livingstone, has left us valuable information bearing on this subject. Speaking of the region of South Central Africa, he states that 'the land now so dry that one might wander in various directions (especially westward to the Kalahari), and perish for lack of the precious fluid as certainly as if he were in the interior of Australia, was once intersected in all directions by flowing streams and great rivers, whose course was mainly south. These river-beds are still called by the natives '*Melaps*' in the south, but in the north '*Wâdies*,' both words meaning the same thing—'river-beds in which no water now ever flows.' To feed these, a vast number of gushing fountains poured forth for ages a perennial supply.* Theobald Fisher, treating of the change of climate in the lands bordering the Mediterranean, shows that the rainfall was once more abundant than at present, especially in Northern Africa.† Zittel and other writers concur in this view; but more than all, the magnificent valleys, with floors overspread by thick deposits of alluvial gravel, bounded by lofty escarpments of limestone, sandstone, or granite, like those of the Zelegah, the Feiran, Es Sheikh, or El 'Ain, now almost waterless—all bear silent testimony to the same great physical change in the climate of the country.

---

* 'Last Journals,' vol. ii., p. 217. There is much more information bearing on this subject, which is of great value.

† 'Studien über das Klima,' etc.; Peterman's 'Mittheilungen,' 1879.

PART V.

CHAPTER I.

ORIGIN OF THE SALTNESS OF THE DEAD SEA.

It has been generally recognised that the waters of lakes which have no outlet ultimately become more or less saline. Of these, the most important in the Old World are the Caspian, the Sea of Aral, Lakes Balkash, Van, Urumiah, and lastly, the Dead Sea, or as it was originally called, 'the Salt Sea.' The Caspian, owing to its great extent and other causes, is but slightly saline ;* but that with which we have here to deal is the most saline of all. It is probable that the water of the ocean itself has become salt owing to the same cause which has produced saltness in the inland lakes, as it may be regarded as a mass of water without an outlet. The cause of the saltness in such lakes I now proceed to explain.

It has been found that the waters of rivers contain, besides matter which is in a state of mechanical suspension, carbonates of lime and magnesia, and saline ingredients in a state of solution ; and as those lakes which have an outlet, such as the Sea of Galilee, part with their waters and saline ingredients as fast as they receive them, the waters of such lakes remain fresh. It is otherwise, however, with regard to lakes which have no outlet. In such cases, the water is evaporated as fast as it is received ; and as the vapour is in a condition of purity, the saline ingredients remain behind. Thus the waters of such a lake tend constantly to increase in saltness, until a state of saturation is attained, when the excess of salt is precipitated, and forms beds at the bottom of the lake. The contrast presented by the waters of the Sea of Galilee on the one hand, and those of the Dead Sea on the other, though both are fed by the same river, is a

---

* Its density, according to H. Rose, being only 1·0013. During the Glacial (or Pluvial) epoch the waters of the Caspian may have been vastly extended, and (as already stated) it had, in all probability, an outlet into the Arctic Ocean. Hence, it is only in very recent times that it has become an inland lake.

striking illustration of the effects resulting from opposite physical conditions. In the former case, the waters are fresh, and abound in fishes and molluscs; in the latter, they are so intensely salt that all animal life is absent.

The increase of saltness in the waters of the Dead Sea has probably been very slow, and dates back from its earliest condition, when its waters stretched for a distance of about 200 miles from north to south. While the uprising of the land, and the sinking down of the Jordan-Arabah depression, were in progress during the Miocene period, some of the waters of the outer ocean, themselves salt, were probably enclosed and retained; but from the occurrence of the shells in the marls in the Arabah Valley, already described (p. 79), it would appear that, when the waters of the great inland lake were at their maximum elevation, they were sufficiently fresh to allow of the presence of molluscous life. This would be during the Pluvial epoch; but at the stage, represented by the salt-beds of the Jebel Usdum, the waters, which were then over 600 feet higher than at present, must have been saturated with chloride of sodium.

(*a*) *Saline and other ingredients of the Dead Sea water.*—The excessive salinity of the waters of the Dead Sea will be recognised from a comparison with those of the Atlantic Ocean. Thus, while the waters of the ocean give 6 lb. of salt, etc., in 100 lb. of water, those of the Dead Sea give 24·57 lb. in the same quantity; but in both cases, the degree of salinity varies with the depth, the waters at the surface being less saline than those near the bottom. The following interesting results are given by M. Lartet, from analyses made by M. Terreil of specimens of the water brought home in sealed glass tubes from the Dead Sea in 1864, and will illustrate this point:

TABLE OF ANALYSES OF THE WATERS TAKEN FROM DIFFERENT POINTS OF THE SURFACE OF THE DEAD SEA, AND AT DIFFERENT DEPTHS (IN 1,000 PARTS).*

DATE, 1864.	POINTS FROM WHICH THE WATERS WERE TAKEN.	DEPTH IN FEET.	SALINE RESIDUE.	WATER.	DENSITY.	CHLORINE.	BROMINE.	SULPHURIC ACID.	CARBONIC ACID.	MAGNESIUM.	SODIUM.	CALCIUM.	POTASSIUM.
March 20	In sea, near Ras Dale	Surface.	27·078	972·922	1·0216	17·628	0·167	0·202	Trace.	4·197	0·885	2·150	0·474
March 24	Lagune, north of J. Usdum	Do.	47·683	952·317	1·0375	29·826	0·835	0·676	Trace.	3·470	7·845	4·481	0·779
April 7	North-east extremity of the Dead Sea	Do.	205·789	794·211	1·1647	126·521	4·568	0·494	Trace.	25·529	22·400	9·094	3·547
March 18	5 miles east of Wâdy Mrabba	66	204·311	795·689	1·1877	145·543	3·204	0·362	Trace.	29·881	13·113	11·472	3·520
March 15	5 miles east of Ras Feschkah	400	262·648	737·352	1·2225	166·340	4·870	0·451	Trace.	41·306	25·071	3·704	3·990
March 18	5 miles north-east from Wâdy Mrabba	1,000	278·135	721·865	1·2533	174·985	7·093	0·523	Trace.	51·428	14·300	17·269	4·386

From the above it will be seen that the greatest proportion of saline residue was in the deepest water, where the density reached 1·253.

If we compare the above composition of Dead Sea water with that of the River Jordan, we shall find that, with the exception of bromine, which has not been recognised,† the waters of this principal tributary of the Dead Sea contain the same ingredients as those of the Dead Sea itself. The following is the analysis of M. Terreil from samples taken from the River Jordan on the 21st April, 1866, about 2 miles from its embouchure :

ANALYSIS OF THE WATER OF THE RIVER JORDAN.

Density, 1·0010.

Saline residue left from one litre (61·028 cubic inches)	= 0·873
Water	= 999·127

Composition.

Chlorine	0·425
Sulphuric acid	0·034
Carbonic acid	Trace.
Soda	0·229
Lime	0·060
Magnesia	0·065
Potash	Trace.
Silica, alumina, and iron	Trace.
Organic matter	Trace.

From the above analyses it will be apparent that the principal ingredients in the waters of the Dead Sea are the chlorides of lime,

---

* Besides the above, traces of silica, alumina, and iron were also present.

† Owing to the restricted quantity of water subjected to analysis.

16

of magnesia of sodium, and of potassium, and in a smaller proportion, sulphates and bromides of the same substances. The absence of carbonates of lime and magnesia is remarkable, as contrasting with the case of waters of rivers draining lands largely formed of limestone rocks.[*]

The quantity of bromine (occurring as bromide of magnesium) in the waters of the Dead Sea has attracted the attention of naturalists. It is known to surpass that in the waters of any other lake yet examined; and may possibly have its origin in the thermal springs originating in the volcanic depths of the interior beneath the bed of the Dead Sea itself, or of its tributaries. It does not, however, pertain to my province to attempt to discuss the origin of these chemical ingredients, and their mutual reactions—questions which have already been so ably handled by Hitchcock, Bischof, Lartet, and others.

(*b*) *Depth of the Dead Sea.*—The floor of the Dead Sea has been sounded on two occasions: first, by the Expedition under Lieutenant Lynch in 1848, and secondly, by that under the Duc de Luynes. In the former case the maximum depth was found to be 1,278 feet; in the latter 1,217 feet, being close approximations to each other. We may therefore affirm that the floor of the lake descends to nearly as great a depth below its surface, as the surface itself below the level of the Mediterranean Sea.

The section given by Lynch indicates that the place of greatest depth lies much nearer the Moabite than the Judæan shore, and the descent from the base of the Moabite escarpment below Jebel Attarus and between the outlets of the Wâdies Mojib and Zerka Maïn, is very steep indeed. The deepest part of the trough seems to lie in a direction running north and south, at a distance of about 2 miles from the eastern banks; and while the ascent towards this bank is rapid, that towards the Judæan shore on the west is comparatively gentle. The line of this deep trough seems exactly to coincide with that of the great Jordan Valley fault. From the bottom of the deeper part, the sounding-line brought up specimens of crystals of salt (sodium-chloride); and it can scarcely be doubted that a bed of this mineral, together with gypsum, is in course of formation over the central portions of the Dead Sea.

---

[*] Thus, according to Bischof, we find the waters of the Rhine, near Basel, to contain of carb. lime, 12·79, and carb. mag. 1·35 in 100,000 parts; those of the Danube at Vienna, 8·37 and 1·50; those of the Thames at Kew, 15·57 and 1·67 respectively. Bischof also gives several sets of analyses of the Dead Sea waters, 'Chem. and Phys. Geol.,' vol. i., p. 92.

CHAPTER II.

RECENT CHANGES OF CLIMATE AND THEIR CAUSES.

HAVING now passed in review the succession of rocks and formations entering into the structure of Arabia Petræa and Palestine, and the causes which have produced the remarkable physical features of this part of the earth's surface, it only remains for me to discuss the changes which have taken place in the climate and vegetation since the Pluvial period ; and to examine the question :—whether there is any prospect of a return to a condition of things as regards the products of these countries more in accordance with the past than what we now see around.

References in the Old Testament Scriptures, as well as the works of ancient historians, would lead us to infer that before the Christian era, the vegetable productions of Palestine and the adjoining districts must have been largely in excess of those of which the region is capable at the present day. A similar conclusion might also be drawn from the former density of the population of those districts of Southern Palestine and the table-land of the Tíh ;—districts which are now all but barren wastes. That this decrease in the productiveness of the land is not confined to Palestine, but extends to the regions of Moab, Edom, the Haurân on the east of the Jordan-Arabah depression, and to the Sinaitic Peninsula, may also be fairly inferred. The Patriarch Job, whether an actual person or a representative character, may be supposed to have inhabited the Plains of Edom, and the graphic description of the products of that region leads us to infer a fruitfulness little compatible with the existing climatic conditions.* The great capital of Idumæa itself, the city of Petra, nestling amongst the almost naked rocks of Arabia Petræa, must have been in the neighbourhood of fields and pastures sufficient to sustain the wants of a

---

* On the other hand, local tradition amongst the Arabs points to the Haurân as the home of the Patriarch ; on this subject consult Oliphant's 'Land of Gilead,' p. 75 *et seq*.

large community. Palestine itself is repeatedly described as a 'land flowing with milk and honey,' and as 'the glory of all lands' in respect of its fruitfulness. Much doubtless was achieved in the way of irrigation, in the regions referred to, by means of aqueducts; but the aqueducts themselves must have been in communication with sources of supply which have almost disappeared.* On the whole, it must be inferred that a change has taken place in the character of the climate, resulting in a diminution of the rainfall, the drying up of some of the streams, and the falling off in the productiveness of the land as compared with that at the time of the settlement of the Israelites, and even later.

On considering this subject, I have come to the conclusion that this change is due to two causes, one cosmical, and beyond man's control ; the other dependent upon human agency, and consequently capable of being rectified by the hand of man himself. We shall consider these two causes in succession.

(*a*) *Cosmical Causes.*—It will be recollected that when treating the subject of the formation of the valleys and ravines it was shown that during the Pluvial period, which immediately preceded the appearance of man, the rain-fall over these regions must have been large, approximating to that of the temperate zone of Western Europe. At this time the now generally dry valleys, both of Sinai and Palestine, were the channels of large streams, while the great Jordan Valley Lake covered with water large tracts of the interior. It is not much to assume that the greater portion of the then land was covered by forests ; and, as Sir J. Hooker has led us to infer, the cedar forests of the Lebanon may then have been nearly connected with those of both Africa and the Western Himalayas. But as time went on, as the snows of the Lebanon disappeared, and as the great lake itself gradually became contracted in its area, a change in the climate began, which must have affected the character of the flora, causing some plants and trees to disappear when the climatic conditions became unfavourable to their habits. The plants suited to temperate climes would have migrated northwards, while those of subtropical regions would have taken their place.

(*b*) *Human Agency.*—At this period primæval man himself invaded

* Léon de Laborde describes the remains of a fine aqueduct near Petra, but the present springs are quite insufficient for its supply.

the region, migrating from the more eastern home of his race, and living by the chase. Monuments of his presence are still found on the plains of Moab in the dolmens, cairns and stone circles which are so abundant east of the Jordan Valley,* and which were, in all probability, equally numerous on the west side previous to the Israelitish occupation. For a long time probably his presence but little affected the vegetation ; but the felling of trees, and the clearing of the ground for the purpose of cultivation, would be the natural consequence of his progress in the arts and civilization.

At the time, therefore, when the ancient inhabitants which were in possession of the country on the invasion of the Israelites had replaced the aboriginal tribes, the country may have to a considerable extent been cleared of its forests. Still, it may be inferred from Bible references that they overspread considerable tracts from which they have now disappeared ; and, in the case of the Sinaitic Peninsula, an inference in the same direction might be drawn from the occurrence of copper-smelting works, carried on by the ancient Egyptians as far back as the reign of Cheops of the fourth or Phiops of the sixth dynasty.† The extreme productiveness of Palestine and its borders about 1,500 B.C. may be referred to the presence of forests over considerable areas, and to the lingering effects of the former climatic conditions, which had not entirely passed away. The cosmical changes were, however, gradually at work, and these were accelerated by the hand of man. The felling of timber appears to have progressed with extraordinary rapidity during the reign of Solomon. The cutting down of cedars and fir-trees in the Lebanon on such a prodigious scale as described in the Book of Kings‡ may be supposed to have been only the chief part of a very general system of disafforesting, which was probably unaccompanied by planting ; for doubtless the timber cut down in the time of the early Kings of Israel, and subsequently, was of natural growth ; and it is not probable that the necessity for keeping up the

---

* It is doubtful whether the 'nawámís' of the Sinaitic Peninsula can be considered as the habitations of primæval man or of later inhabitants. See Wilson's Notes in 'Ordnance Survey of Sinai,' part ii., p. 194. Mr. Laurence Oliphant has recently discovered the overturned blocks of a Dolmen in Judæa, between Jericho and Nablus. ('Quarterly Statement,' July, 1885, p. 181).

† According to Dr. S. Birch, 'Ordnance Survey of Sinai,' ch. vii., p. 167.

‡ 1 Kings v. 6, *et seq.*

supply by planting would have been recognised at this epoch.   The effect, however, on the climate could not fail to be marked ; and we may suppose that the rainfall must have lessened considerably during the latter period of Jewish history.   The injurious influence upon agriculture consequent on the lessening of the rain may have been largely mitigated by an extensive and systematic practice of irrigation, by which the waters of the springs were carried long distances for moistening the ground.   But once these fell out of repair, as they must have done in the calamitous times succeeding the fall of Jerusalem, the land would have reverted to a con-dition of sterility ; while the rains falling on the unprotected slopes would wash the soil down into the valleys, and leave the sides of the valleys, at least over the tableland of Judæa, bare and rocky as they are at this day.*

Palestine and Arabia Petræa suffer from the same cause which has rendered barren and desolate large tracts of Northern Africa, Spain, and Portugal, Southern and Central Russia, and threatens to reduce to a similar condition extensive tracts of the continent of the New World. The disafforesting of such a country as Palestine and Syria cannot but have proved prejudicial to its productiveness.   It operates in two ways. First, the ground becomes unduly heated by exposure to the direct rays of the sun, and the heat being radiated into the air disperses the clouds.   In consequence of this, the rain only falls when there is some extraordinary electrical disturbance, producing thunderstorms which break in heavy falls of rain at distant intervals throughout the year : and secondly, the rainfall, instead of being gently distributed over the ground by the agency of the foliage, descends directly on the surface, and sweeping down into

---

* I have been favoured with the following note by Colonel Sir Charles Wilson : ' I think you have rather over-estimated the change of climate and diminution of rainfall during historic times.   My own opinion is that there has been but slight change since the conquest by Joshua, either in climate or rainfall ; the great difference is that the disappearance of the forests and want of proper cultivation has allowed the rainfall to run off rapidly instead of gently ; this, of course, has had a great effect on the formation of cloud, which at one time moderated the temperature.   The great importance attached to springs in the Bible, the allusions to drought and periodic rainfall, as well as the extraordinary number of cisterns in all parts of the country, and the extensive remains of aqueducts, seem to show that there was always a deficiency of rainfall.   Add to this the conditions of rainfall were always the same, viz., the warm westerly winds, charged with moisture from the Mediterranean, striking the hills of Palestine.   Of course this is matter of opinion, and may be argued both ways."

the valleys with impetuosity, carries away the soil and lays bare the rock. In this way we may account for the rocky and naked character of the valley sides and faces of the hills throughout the central table-land of Palestine and in Syria, and for the extraordinary depth of the loam and soil in the flat lands and broad valleys.*

(*c*) *Mode of Restoration.*—If, therefore, I were asked what course ought to be adopted in order to restore to Palestine its pristine fruitfulness, I would reply: Cultivate the extensive plains of Philistia and Southern Judæa, and plant the hills and valley sides, not only with the vine and the olive, but with forest-trees. For such a beneficial undertaking, Palestine presents special advantages. Owing to its peculiar physical conditions, the height of its elevations, the depth of its depressions, it seems adapted for almost every variety of vegetable product. Tropical plants find a fitting climate in the plains of the Jordan and Salt Sea, while the table-land and northern portions are adapted for oaks, firs, cedars, and forest trees of Europe. The eucalyptus, planted extensively over the plain of the Ghor, would deprive it of its malaria. The slopes and deep glens would offer a fitting habitat for the box, the maple, the walnut, and other trees producing ornamental timber. In fact, Palestine might become what it once was, a land of rivers and fountains of waters ; 'the wilderness be converted into a fruitful field, and the fruitful field be counted for a forest.'

---

* That the process of disafforesting in adjoining districts is still going on may be gathered from the fact that, under the dominion of Muhammed Ali of Egypt, vast quantities of timber were annually cut down upon the hills behind Iskanderûn, and sent to Egypt. These forests are mostly of oak and pine, with some beech and linden. In 1837 the timber of about sixty thousand large trees was shipped to Alexandria, two-thirds of which was for ship-building (Browning's ' Report on Syria,' p. 11 *et seq.*). The Taurus and Amanus ranges have supplied timber to Egypt from the earliest times. Cleopatra got her timber from the former. These forests are now being ruined by stripping the trees of their bark for export.

APPENDIX A.

DOMESTIC REMEDIES OF THE ARABS OF THE DESERT :
WITH ETHNOLOGICAL NOTES, ETC.

BY

E. GORDON HULL, M.D., B.A.,
Dublin.

During the autumn and winter of 1883-4, while acting as honorary medical officer and assistant-geologist to the Expedition sent out by the Palestine Exploration Fund to Arabia Petræa and the Arabah Valley, I had an opportunity of making some notes on the common ailments and remedies of the Bedawin, and also some observations on their physical condition, which I will state as shortly as possible.

Physically, they are a small race, but their limbs and bodies are well formed and proportioned, and they are capable of enduring great fatigue on a diet consisting principally of boiled rice and butter, with unleavened bread, coffee, dates, and water. We came across two principal tribes, those of the Sinaitic Peninsula, and those who live east of the Arabah Valley. The Towara, who live between the Gulfs of Suez and Arabah, are more civilized than their eastern brethren, and indulge in a little agriculture, but they live chiefly by tending sheep and goats.

I measured twenty-six adult males, picked men of the tribe, taking three measurements, that is, height, chest round nipple, and length of right arm from acromion to top of middle finger. The average height was 5 feet 4½ inches; average chest-measurement, 31 inches; average length of right arm, 27·6 inches. The maximum chest-measurement in a man of 5 feet 11 inches was only 34¼ inches, and the minimum in one of 5 feet was 30 inches. Yet, with such insignificant chests, they were splendid pedestrians and mountaineers, and did their day's march without a murmur. Certainly, they were all in very good condition, for I do not suppose there was an ounce of fat among the whole tribe. The next tribe with whom we came into contact were even of smaller make, averaging only 5 feet 2 inches in height. Their muscles, especially of the upper extremities, were very poorly developed, while they nearly all, except the Sheik, exhibited marks of inferior intelligence; about five or six out of our twenty men were decidedly half-witted, and all of them had the habit, common among such people, of repeating over and over again everything that is said to them, or that they say to one another. They appeared half-starved, and used to chew continuously the dried beans provided for the camels. Like most wild animals, they have splendid teeth, very firmly fixed in the jaw, and their sight is apparently keen. They nearly all turn in their toes when walking.

Owing, no doubt, to their habits, the Arabs seem to be most subject to the diseases due to exposure; but, as far as I could judge, these diseases have a tendency to the chronic or subacute form, rather than the acute. Two I particularly noticed as almost universal, that is, chronic bronchitis, of a dry kind, and without emphysema; and chronic articular rheumatism.

The first cannot fail to force itself on every traveller's attention, as it gives rise to a peculiarly irritating paroxysmal cough, rather canine in character, which, as the Arabs sat round our tents at night, often disturbed our slumbers. The second, that is, rheumatism, I noticed when taking the measurements of a series of the men; nearly all their shoulder-joints creaked and groaned as they raised them; and this will account for the curious inability of the Arabs to move about or do any work in the morning before they are 'thawed' and rendered supple, either by fire or by sun. It seems, at such a time, as if all their joints were temporarily ankylosed, so stiff and unpliable are they. Ophthalmia is common; it appears to begin usually as a purulent conjunctivitis, but is very liable to attack the cornea; and, in a great many cases, it proved to be really associated with granular lids. A great number of the Bedawin of the Peninsula had corneal opacities; but, amongst the wilder tribes of the east of the Arabah, I noticed little or none, neither did any that I came across suffer from conjunctivitis or ophthalmia of any kind; and, partly on this account, and partly from the well known contagious nature of the granular lid, and the habits of the Arabs of wiping anything and everything, their own eyes and their children's, etc., on their never washed calico shirts, I am rather disposed to look upon their ophthalmia as a contagious disease, acquired from contact with the fellaheen of the cities of Egypt and Palestine, among whom it is almost universal. And in this opinion I am strengthened by the observations of Mr. Merrill, the American Consul at Jerusalem, who told me that there was no ophthalmia among the great tribes east of Jordan. Another argument against the opinion that their ophthalmia is due to the heat and sand of the desert is that, while ophthalmia is prevalent among the Arabs inhabiting the comparatively elevated plateau and stony valleys of the Sinaitic Peninsula, those who live in the great deserts of Arabia Magna do not seem to suffer from it at all; and, at Akabah again, where a number of low-class Arabs are crowded together in huts, ophthalmia was very prevalent. I believe that in this, as in most other cases, it is originally due to dirt and overcrowding, and is then spread by contagion.

Constipation is universal, and sometimes appears to simulate a false dysentery, with great pain in the bowels, which is relieved at once by a purge. They call a pain in the abdomen, or a colic, a 'heartache,' and I have removed many a 'heartache' with a gamboge pill. Their habit of over-eating when they get the chance, alternating with more or less prolonged periods of semi-starvation, will account for the prevalence of this complaint.

Skin-disease is not so frequent as one would expect from their un-cleanly habits, if we except phtheiriasis; however, I saw two cases of impetigo, and I was given an account of a kind of moist purulent eruption, which, in the spring-time of the year, is apt to attack both men and camels, especially the latter. The hair falls off the places attacked, and the seat of the disease becomes covered with scabs; it also has a very unpleasant smell. They call it 'jarrub,' and believe it to be catching. Sulphur, externally and internally, made into a paste with butter, is the native remedy for this disease.

In certain districts, notably Akabah, they suffer from ague, but this disease is not common among the Arabs proper. The late Professor Palmer states that 'they are sometimes visited by an epidemic, not cholera, probably the plague, which they call "the yellow pest." It comes with the hot winds, and strikes them down suddenly in the midst of their occupation; but it is said never to attack the country of our Lord Moses, where grow the shiah and the myrrh, that is, the elevated granite-region about Mount Sinai.'

It will be seen from the above that the diseases to which the Arabs are subject are few; and if we except those infectious diseases which come to them from without, such as cholera and small-pox, they appear to be a very healthy people.

The remedies they are in the habit of using are not many in number and are usually derived from those plants which are most widely distributed. As a diuretic, that is, for pains in the back, and gravel, they use the retem, or broom (*Retama retam*) making a decoction of the top shoots in hot water, and drinking it; they say it is also purgative. This shrub, which provides them with fuel and their camels with a scanty nourishment, is almost universal; we saw it in flower on the way to Petra, and the inflorescence, which is purple and white, gives out an exceedingly sweet perfume. It has a very bitter taste.

Several species of wild melon, of the family *Bryoniæ*, allied to the elaterium (which also grows in these parts), are in common use as purgatives; the native method of using them is ingenious. A fruit is split into halves, the seeds scooped out, and the two cavities filled with milk; after allowing it to stand for some time, the liquid, which has absorbed some of the active principle of the plant, is drunk off. A milder remedy is camel's milk, which appears, under some circumstances, to be purgative to the Arabs.

The order *Compositæ* furnishes several medicinal herbs of which the Arabs make use. The *Santolina fragrantissima*, a graceful plant of a sage-green colour, bitter taste, and strong fragrant smell, furnishes them, in the form of an infusion, with a carminative, good for colic and all painful affections of the abdomen. In the bazaars of Cairo, the fragrant dried heads are sold for the same purposes as camomile. I was told that there are no snakes in the district where the plant grows; and the natives believe that the smell of the plant is sufficient to drive reptiles from a house, and it is used for this purpose in Cairo and other towns.

Another plant of the same order is an *Artemisia*, or wormwood, with a very strong aromatic odour and bitter taste. The fellaheen use it to put in their bedding to drive away vermin. This use of the plant appears to be very universally known; witness the old English rhyme—

'When wormwood hath seed, get a handful or twain,
To save against March, to make flea to refrain:
Where chamber is swept, and wormwood is strewn,
No flea, for his life, dare bide, or be known!'

From the seed of some of the kinds of *Artemisia*, which grow in these parts, santonine appears to be obtained. In the wilderness of Judæa, near to Beersheba, we found a pretty little *Calendula*, or marigold, very common. It became extremely abundant along the Mediterranean sea-board, and is used by the natives as a sort of tea for flatulence and pain in the abdomen. Knowing how largely a liniment derived from this plant was advertised by homœopaths, I tried to find out if they used it as an external application, but they did not know of its virtues as such. One of the commonest desert-plants is the zygophy̆llum, so called from the leaves being composed of short succulent jointed segments; these, bruised in water, form a mucilaginous liquid, of which the Arabs are very fond as an application for sore eyes. It has an exceedingly nauseous taste, but this fact only appears to commend it to the notice of the camel, who devours it greedily. A curious tropical plant, which we found in the Ghor, at the south end of the Dead Sea, is the osher (*Callotropis gigantea*), a large tree-like asclepiad, containing simply enormous quantities of milky acrid juice. Its properties are stated by Endlicher to be powerfully purgative and emetic; but the natives use it to give to women whose milk is scanty, probably in accordance with the doctrine of signatories. Here, also, grows the castor-oil plant, but its virtues are unknown to the natives. A very striking plant, which, perhaps, I should have mentioned before, and which often hangs in graceful dark-green festoons from the granite walls of the gorges of Arabia Petræa, is the caper plant (*Capparis spinosa*). The natives are very fond of the fruit, which has a warm aromatic taste, and they stroke the region of the epigastrium appreciatively after eating one or two. The cortex of the root is said to be aperient and diuretic. Another fairly common plant is a hyoscyamus, called by the natives sekharan, with fleshy leaves and purple flowers. The dried leaves are used by the natives to smoke, and produce a kind of intoxication or delirium; and an infusion of the fresh leaves possesses strong narcotic properties. It is nearly allied to the mandragora, which becomes common on the limestone downs in the south of Judæa. The Arabs are extraordinarily susceptible to narcotics. Our tobacco they cannot smoke at all; a few whiffs make them giddy, and give them a headache; even a 'Richmond Gem' cigarette is too much for them. Only two mineral substances appear to be regarded by the Bedawin as medicinal. One of them, sulphur, I have already mentioned; the other is a kind of common red coral, found on the shores of the Red Sea and

Mediterranean, and sold in the bazaar at Gaza. As far as I could gather, they only use this as a charm.

In conclusion, I may mention that the few drugs which are really useful on an expedition of this sort can be carried in a small tin box a few inches square. And a little doctoring is greatly appreciated by the natives. Purgative pills, quinine, some preparation of opium, a bottle of nitrate of silver and atropine drops for sore eyes, and some strong ammonia and powdered ipecacuanha for bites and stings, will go a long way. I took a good deal more than this, as I had a large party to 'keep in repair,' but the above list contains the drugs most generally useful.

APPENDIX B.

NOTES ON THE SPECIMENS OF ROCK COLLECTED BY
PROFESSOR HULL, F.R.S., DURING THE EXPEDITION
TO ARABIA FOR THE PALESTINE EXPLORATION
COMMITTEE.

BY

F. W. RUDLER, F.G.S.,
Curator of the Museum of Practical Geology, London.

1. *Syenitic Gneiss, from Summit of Jebel Musa.*—A crystallo-granular, slightly foliated rock, composed of opaque white and pale pink felspars, associated with quartz and with scattered patches of dark-green hornblende, which impart a speckled appearance to the rock. Under the microscope the felspars appear as ill-defined crystals, more or less altered, and granular in texture. Some of them are interlaminated so as to form a kind of microperthitite. The quartz occurs as large irregular crystalline grains, with numerous fluid enclosures, which, in some cases, are disposed roughly parallel to the outlines of the mass. The large crystals and crystalloids of felspar and quartz are embedded in a mass of fine-grained texture, composed of the same minerals. Here and there the twin-striation of plagioclase may be indistinctly seen, while in some of the crystals fine cross-hatching may be detected ; probably the felspathic components of the rock are orthoclase, microcline, and oligoclase. The hornblende forms granular masses of dark-green colour, with scarcely any characteristic cleavage. It is associated with a chloritic mineral. Apatite is present in prismatic crystals piercing the hornblende, and elsewhere ; while a little ferrite is scattered through the rock, as a product of decomposi-

tion, and is especially evident along the fissures which traverse the specimen.

2. *Red Syenitic Granite, from Flanks of Jebel Musa.*—Consisting mainly of flesh-coloured orthoclase, vitreous quartz, and black hornblende. The felspar appears under the microscope in large crystals, more or less nebulous from incipient decomposition, and the included granular matter exhibits a tendency to arrange itself in patches with a rough parallelism. Peroxide of iron is disseminated, as a reddish dust, through much of the felspar, conferring upon it the characteristic flesh-colour. Triclinic felspars, including microcline, may be detected. The quartz forms large clear rounded masses, with numerous pores arranged in linear series. The hornblende appears in ill-defined masses of dark-brown colour. A little epidote is present, and a good deal of ferrite. This specimen on the whole resembles the last in its mineral constituents, but differs from it in presenting a coarser granitoid texture and a redder colour.

3. *Fine-grained Pink Granite, from a Dyke in Jebel Watiyeh.*—This rock is a crystallo-granular aggregate of red and white felspars and quartz, with here and there a flake of dark-brown mica. Viewed under the microscope, the felspars appear cloudy, from having suffered partial kaolinization, and in many cases the outlines of the crystals are without sharp definition. There is a marked tendency for the included granular matter to arrange itself in concentric zones parallel to the edges of the crystals. The plagioclastic felspar shows characteristic polysynthetic twinning, and the extinction angles of the component laminæ, taken in relation to the trace of the twin-section, give values ranging from 5° to 18°. Some of the felspar may be albite. The quartz is in clear grains, many of which are elongated, and all are highly charged with fluid enclosures. There is scarcely any mica visible in the section, but a few minute flakes of biotite, mostly altered to chlorite, and some folia of muscovite are sparsely distributed through the rock.

4. *Coarse-grained Red Granite, from Wády es Sheikh.*—This specimen is composed of pink felspar, with a white felspar in subordinate quantity, colourless quartz and dark-brown mica; while chlorite and epidote are present as secondary products. Studied microscopically, the felspars are seen to form large irregular nebulous crystals, some of which show broad ill-defined twin lamellæ, with extinction angles fairly agreeing with those of microcline (above 15°). The quartz presents a dusty appearance, due to numerous pores. There has been much biotite, or dark magnesian mica,

in the rock ; but most of this has been converted into a chloritic product, which is associated, as usual, with magnetite. In one case the epigenic chlorite presents a sharp hexagonal outline, having preserved the form of the original biotite. The rock-specimen is friable, and the microscopic preparation is in three small fragments. One of these pieces contains a good deal of epidote, in characteristic groups of crystals, of pale yellow colour, with high refractive index, and giving vivid colours between crossed Nicols. Apatite occurs in rather stout crystals, and peroxide of iron, as a product of decomposition, is disseminated through the rock.

5. *Coarse-grained Pink Granite, from Wâdy es Sheikh.*—Containing two felspars—one of flesh-colour, the other opaque white—associated with vitreous quartz and numerous folia of biotite. The felspars are seen in the microscopic section to be highly charged with included matter which imparts a dense turbidity to some of the crystals. There is a decided tendency to a zonal arrangement of the enclosed granules. Indistinct twin-striation shows that some of the felspar is triclinic, oligoclase being probably associated with orthoclase. The quartz is much fissured, and ferric oxide occupies many of the cracks, while the usual cavities containing liquid give a dusty appearance to the mineral. The biotite is of the ordinary type, occurring in laminæ, which, when cut parallel to the principal axis, offer a striated appearance, due to the edges of the basal cleavage planes. The mica is strongly dichroic, varying in tint from brown to black, and in some cases it has degenerated into green chloritic products, associated with a little magnetite. A crystal displaying an acute rhomboidal section may be sphene, and a few crystals of apatite are present. This section is rather thick, and the specimen, like most of the others, has evidently suffered from weathering.

6. *Diabase, occurring as a Dyke in Wâdy es Sheikh.*—This is a dark-coloured, fine-grained, basalt-like rock, with disseminated crystals of iron pyrites. Microscopic examination shows that the felspar presents lath-shaped sections, with ill-defined outlines, often ragged at the ends. They are mostly binary twins, probably of labradorite, but too much altered for determination. Here and there a stout prismatic crystal may be seen, with indistinct zonal structure. The pyroxenic constituent occurs in small yellowish-green grains and indistinct crystals. Numerous opaque grains of iron ore are present, and the rock contains much calcite, which is evidently a product of decomposition. The section shows

the junction of two rocks, one coarser in grain than the other, and with larger felspar crystals porphyritically embedded in the ground-mass.

7. *Porphyrite, forming a Dyke on the East side of Wâdy el Arabah.*—This rock would probably be termed by some petrologists a melaphyre, and by others a diabase-porphyrite. It consists of a dark fine-grained base with pinkish felspar, disseminated in the form of crystals and irregular patches, and associated with epidote. Under the microscope the felspar appears in irregular cloudy crystals ; but notwithstanding the evident alteration to which they have been subjected, many still exhibit indistinctly the twin striation of plagioclase. The large felspars are embedded in a ground-mass made up of small lath-shaped crystals of felspar, with viridite and opaque granules of an oxide of iron. Epidote is abundant, occurring in congeries of very pale greenish-yellow crystals, sharply defined and full of cleavage-fissures. Calcite is freely dissemi-nated through the rock, and a few flakes of red oxide of iron are present.

8. *Pink Granite, from East side of Edomite Mountains.*—Composed mainly of red felspar and quartz, with a few dark specks, which in this specimen, representing an altered condition of the rock, seem to be chiefly iron ore. The felspar predominates, and appears under the microscope in ill-defined cloudy crystals, some of which enclose a reddish dust of peroxide of iron. Most of the felspar appears to be orthoclastic ; but there is also present a fair proportion of plagioclase, which, from the extinction-angles, seems to be oligoclase, associated with microcline. The quartz presents a reticulation of pores. A single flake of biotite may be detected in the section, and several grains of epidote are visible. Apatite occurs .in long clear prismatic crystals, embedded in the quartz. The opaque black mineral, which is seen in several parts of the section, is probably titaniferous iron ore, inasmuch as it is associated with an alteration-product resembling leucoxene. Red oxide of iron occurs in fissures traversing the specimen.

9. *Porphyrite from a Dyke in Wâdy el Arabah.*—A fine-grained dark-coloured rock, with green epidotic patches and a rusty exterior. This specimen has evidently suffered much alteration. It contains in its present state a great deal of epidote, which appears to have partly taken the place of large felspar-crystals. The ground-mass is made up of a confused congeries of ragged lath-shaped felspars, associated with green

18

products of decomposition and granules of magnetite. Slender prisms of apatite are numerous, and much brown oxide of iron is present.*

10. *Spherulitic Quartz Felsite, from a Dyke in Wâdy el Arabah.*—A porphyritic rock with fine-grained, chocolate-coloured ground-mass, containing rounded crystals or 'blebs' of vitreous quartz, with opaque white crystalline patches of altered felspar. Viewed in the microscope, the quartz presents irregular outlines, being broken into locally by embayments of the ground-mass, of which it also contains apparent inclusions. There are numerous cavities with liquid, disposed in lines traversing the quartz. The felspar, apparently orthoclase, occurs in ill-defined crystals which have suffered much alteration. Calcite and epidote are present as products of decomposition. The micro-crystalline ground-mass is studded with spherulites and indistinct spherulitic patches, with radial structure, presenting between crossed Nicols the common Payne's grey colour, with ill-defined black crosses. Some of the spherulites have a nucleus of quartz, and many of the larger crystals of quartz and felspar are bordered by a fringe of radiating fibres disposed in tufts. Most of the spherulitic areas contain microliths, and in many cases these seem to be felspathic, and are forked at each end. Ferrite is freely disseminated through the rock, and no doubt confers upon the matrix its general chocolate or liver-colour.

11. *Pink Granite, from East side of Wâdy Arabah.*—A coarse-grained aggregate, mainly of pink felspar and quartz, with scattered patches of limonite. No mica is visible, but the specimen is evidently much altered, and possibly the oxide of iron may represent what was originally a dark mica. Two felspars are present. The dominant species appears to be orthoclase in large irregular crystals, some of which are twinned, and penetrated by laminations of another felspar, probably microcline, while others contain calcite. The orthoclase is associated with plagioclase in small crystals and grains, much altered, yet retaining evidence of twin-lamellation. The quartz retains its normal characters, being present in fairly clear crystalloids, containing lines of pores and a few crystals of apatite and other endomorphs. There is much ferrite along the fissures of the specimen, and some opaque iron-ore which is probably magnetite.

12. *Quartz Diabase, from East side of Wâdy el Arabah.*—This is an altered eruptive rock which would formerly have been called 'greenstone' or melaphyre. It presents a dark-greenish colour and fine-grained texture,

---

* This is very similar to specimen No. 7.

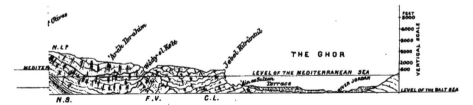

V. Felstone.

SDUM.

TABLE LAND OF MOAB

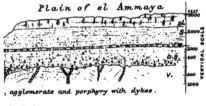

, agglomerate and porphyry with dykes.

nic Beds.

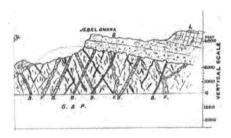

:lstone Porphyry dykes.

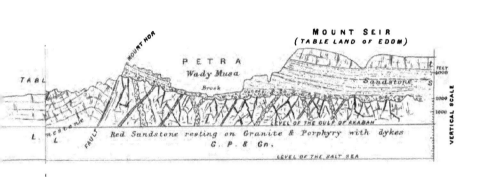

MOUNT SEIR
(TABLE LAND OF EDOM)

PETRA
Wady Musa
Brook

MOUNT HOR

TABL

Sandstone

LEVEL OF THE GULF OF AKABAH

FEET
4000

3000

1000

VERTICAL SCALE

L

FAULT

Red Sandstone resting on Granite & Porphyry with dykes
G. P. & Gn.

LEVEL OF THE SALT SEA

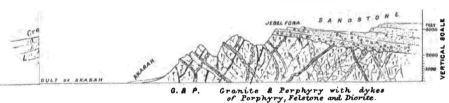

MOUNTAINS OF EDOM

SANDSTONE

JEBEL FORA

FEET
4000

2000

1000

VERTICAL SCALE

Cre

AKABAH

GULF OF AKABAH

G. & P.   Granite & Porphyry with dykes
of Porphyry, Felstone and Diorite.

Schist.   G. Granite.   P. Porphyry &c.

Edwd Weller, lith.

IH.

Gn. & St. Gneiss & Schists &c.

H.

North

JEBEL EJMEH

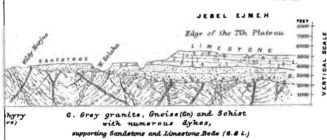

G. Grey granite, Gneiss (Gn) and Schist
with numerous dykes,
supporting Sandstone and Limestone Beds (S. & L.)

Edw⁴ Weller lith.

with patches and strings of epidote and nests of free quartz. Under the microscope it appears as a dense network of irregular lath-shaped microliths of plagioclase, with intervening viridite of pale-green colour, and grains of magnetite. Much epidote is present in the common form of clusters of yellowish-green gum-like grains and groups of characteristic crystals. The quartz is in clear masses, consisting of aggregated crystalloids. Hematite occurs in bright red translucent six-sided crystals, and grains of an opaque oxide of iron are scattered through the rock.

13. *Hornblende-augite Andesite, from Jebel esh Shomrah.*—A coarsely crystalline dark-coloured rock, showing macroscopically crystals of black hornblende, with a little felspar. Some reddish patches, somewhat like garnet, seem to be composed of an altered felspathic mineral. Studied microscopically, the hornblende appears as large greenish-brown, slightly dichroic crystals, with rather indistinct outlines. They are much shattered and fissured. Sections transverse to the principal axis show in places the characteristic prismatic cleavage with angles of 124°. The twin structure is seen in some of the sections when examined between crossed Nicols. The carious outlines of the hornblende crystals are mostly surrounded by a border of darker greenish granules. The augite forms large pellucid crystals, and some of the sections cut transversely to the principal axis show octagonal outlines and characteristic prismatic cleavages with approximately rectangular intersections. A chloritic mineral is developed along some of the fissures in the augite, and the habit of some of the crystals is suggestive of that of olivine. The ground-mass is of pale-brown colour, crowded with felspathic microliths, associated with a pale-green chloritic material and with grains of magnetite. None of the ground-mass seems to be quite isotropic. Felspar occurs not only in the small rod-like crystals of the matrix, but in larger crystals with zonal structure : these, however, are rare. Among the products of decomposition are chlorite, or an allied mineral, forming large areas with irregular outlines and a slightly fibrous structure ; serpentine in olive-green patches ; a few grains of epidote ; calcite in crystalline masses ; and peroxide of iron, included within some of the crystals, and also disseminated in flakes throughout the rock. By some petrologists this rock would probably be classed as a hornblende-porphyrite.

The writer desires to acknowledge the kind assistance of Mr. F. Rutley, F.G.S., in determining some of the minerals in the rocks described above.

INDEX.

THE END.

BILLING AND SONS, PRINTERS, GUILDFORD.